Solving Problems in Food Engineering

Stavros Yanniotis, Ph.D.
Author

Solving Problems in Food Engineering

 Springer

Stavros Yanniotis, Ph.D.
Department of Food Science and Technology
Agricultural University of Athens
Athens, Greece

Additional material to this book can be downloaded from http://extras.springer.com.

ISBN 978-0-387-73514-6   ISBN 978-0-387-69358-3 (eBook)

Library of Congress Control Number: 2007939831

Printed on acid-free paper

9 8 7 6 5 4 3 2 1

springer.com

*"Tell me and I will listen,
Show me and I will understand
Involve me and I will learn"*

Ancient Chinese Proverb

Preface

Food engineering is usually a difficult discipline for food science students because they are more used to qualitative rather than to quantitative descriptions of food processing operations. Food engineering requires understanding of the basic principles of fluid flow, heat transfer, and mass transfer phenomena and application of these principles to unit operations which are frequently used in food processing, e.g., evaporation, drying, thermal processing, cooling and freezing, etc. The most difficult part of a course in food engineering is often considered the solution of problems. This book is intended to be a step-by-step workbook that will help the students to practice solving food engineering problems. It presumes that the students have already studied the theory of each subject from their textbook.

The book deals with problems in fluid flow, heat transfer, mass transfer, and the most common unit operations that find applications in food processing, i.e., thermal processing, cooling and freezing, evaporation, psychometrics, and drying. The book includes 1) theoretical questions in the form "true" or "false" which will help the students quickly review the subject that follows (the answers to these questions are given in the Appendix); 2) solved problems; 3) semi-solved problems; and 4) problems solved using a computer. With the semi-solved problems the students are guided through the solution. The main steps are given, but the students will have to fill in the blank points. With this technique, food science students can practice on and solve relatively difficult food engineering problems. Some of the problems are elementary, but problems of increasing difficulty follow, so that the book will be useful to food science students and even to food engineering students.

A CD is supplied with the book which contains solutions of problems that require the use of a computer, e.g., transient heat and mass transfer problems, simulation of a multiple effect evaporator, freezing of a 2-D solid, drying, and others. The objectives for including solved computer problems are 1) to give the students the opportunity to run such programs and see the effect of operating and design variables on the process; and 2) to encourage the students to use computers to solve food engineering problems. Since all the programs in this CD are open code programs, the students can see all the equations and the logic behind the calculations. They are encouraged to see how the programs work and try to write their own programs for similar problems. Since food science students feel more comfortable with spreadsheet programs than with programming languages, which engineering students are more familiar with, all the problems that need a computer have EXCEL® spreadsheet solutions.

I introduce the idea of a digital SWITCH to start and stop the programs when the problem is solved by iteration. With the digital SWITCH, we can stop and restart each program at will. When the SWITCH is turned off the program is not running, so that we can change the values of the input variables. Every time we restart the program by turning the SWITCH on, all calculations start from the beginning. Thus it is easy to change the initial values of the input variables and study the effect of processing and design parameters. In the effort to make things as simple as possible, some of the spreadsheet programs may not operate on some sets of parameters. In such cases, it may be necessary to restart the program with a different set of parameters.

I am grateful to Dr H. Schwartzberg, who read the manuscripts and made helpful suggestions. I will also be grateful to readers who may have useful suggestions, or who point out errors or omissions which obviously have slipped from my attention at this point.

Athens   Stavros Yanniotis
May 2007

Contents

Chapter 1
Conversion of Units

Table 1.1 Basic units

	Time	Length	Mass	Force	Temperature
SI	s	m	kg	–	K, ^{0}C
CGS	s	cm	g	–	K, ^{0}C
US Engineering	s	ft	lb_m	lb_f	^{0}R, ^{0}F

Table 1.2 Derived units

	SI	US Engineering
Force	N (1 N = 1 kg m/s^2)	–
Energy	J (1 J = 1 kg m^2/s^2)	Btu
Power	W (1 W = 1 J/s)	HP, PS
Area	m^2	ft^2
Volume	m^3 (1m^3 = 1000 l)	ft^3
Density	kg/m^3	lb_m/ft^3
Velocity	m/s	ft/s
Pressure	Pa (1 Pa = 1 N/m^2)	psi = lb_f/in^2
	bar (1 bar = 10^5 Pa)	
	torr (1 torr = 1 mmHg)	
	atm (1 atm = 101325 Pa)	

Table 1.3 Conversion factors

1 ft = 12 in = 0.3048 m	^{0}F = 32+1.8* ^{0}C
1 in = 2.54 cm	^{0}C = (^{0}F-32)/1.8
1 US gallon = 3.7854 l	^{0}R = 460 + ^{0}F
1 lb_m = 0.4536 kg	K = 273.15 + ^{0}C
1 lb_f = 4.4482 N	
1 psi = 6894.76 Pa	Δ^0C = Δ^0F/1.8
1 HP = 745.7 W	Δ^0C = ΔK
1 Btu = 1055.06 J = 0.25216 kcal	Δ^0F = Δ^0R
1kWh = 3600 kJ	

S. Yanniotis, *Solving Problems in Food Engineering.*
© Springer 2008

1

Examples

Example 1.1

Convert 100 Btu/h ft^{2o}F to kW/m^{2o}C

Solution

$$100\frac{\text{Btu}}{\text{h ft}^2\,{}^\circ\text{F}}=100\frac{\text{Btu}}{\text{h ft}^2\,{}^\circ\text{F}}\cdot\frac{1055.06\,\text{J}}{1\,\text{Btu}}\cdot\frac{1\,\text{kJ}}{1000\,\text{J}}\cdot\frac{1\,\text{h}}{3600\,s}\cdot\frac{1\text{ft}^2}{(0.3048\,\text{m})^2}$$

$$\cdot\frac{1.8\,{}^\circ\text{F}}{1\,{}^\circ\text{C}}\cdot\frac{1\,\text{kW}}{1\,\text{kJ/s}}=0.5678\frac{\text{kW}}{\text{m}^2\,{}^\circ\text{C}}$$

Example 1.2

Convert 100 lb mol/h ft^2 to kg mol/s m^2

Solution

$$100\frac{\text{lb mol}}{\text{h ft}^2}=100\frac{\text{lbmol}}{\text{h ft}^2}\cdot\frac{0.4536\,\text{kg mol}}{\text{lb mol}}\cdot\frac{1\,\text{h}}{3600\,\text{s}}\cdot\frac{1\,\text{ft}^2}{(0.3048\,\text{m})^2}=0.1356\frac{\text{kg mol}}{\text{s m}^2}$$

Example 1.3

Convert 0.5 lb$_f$ s/ft^2 to Pas

Solution

$$0.5\frac{\text{lb}_f\,\text{s}}{\text{ft}^2}=0.5\frac{\text{lb}_f\,\text{s}}{\text{ft}^2}\cdot\frac{4.4482\,\text{N}}{\text{lb}_f}\cdot\frac{1\,\text{ft}^2}{(0.3048\,\text{m})^2}\cdot\frac{1\,\text{Pa}}{(1\ \text{N/m}^2)}=23.94\,\text{Pa s}$$

Exercises

Exercise 1.1

Make the following conversions:
1) 10 ft lb$_f$/lb$_m$ to J/kg, 2) 0.5 Btu/lb$_m$°F to J/kg°C, 3) 32.174 lb$_m$ft/lbfs2 to kgm/Ns2, 4) 1000 lbmft /s2 to N, 5) 10 kcal/min ft °F to W/mK, 6) 30 psia to atm, 7) 0.002 kg/ms to lb$_m$ft s, 8) 5 lb mol/h ft^{2}mol frac to kg mol/s m^2 mol frac, 9) 1.987 Btu/lbmol °R to cal/gmol K, 10) 10.731 ft^{3}lb$_f$/in^{2}lbmol °R to J/kgmol K

Solution

1) $10\dfrac{ft\,lb_f}{lb_m} = 10\,\dfrac{ft\,lb_f}{lb_m} \cdot \dfrac{\text{...............m}}{ft} \cdot \dfrac{\text{...............N}}{1\,lb_f} \cdot \dfrac{\text{...............}lb_m}{0.4536\,kg} \cdot \dfrac{\text{............}J}{m\,N}$

$= 29.89\dfrac{J}{kg}$

2) $0.5\dfrac{Btu}{lb_m\,°F} = 0.5\,\dfrac{Btu}{lb_m\,°F} \cdot \dfrac{1055.06\,J}{\text{..........}} \cdot \dfrac{\text{............}}{\text{............}} \cdot \dfrac{1.8°F}{1°C} = 2094.4\,\dfrac{J}{kg\,°C}$

3) $32.174\dfrac{lb_m\,ft}{lb_f\,s^2} = 32.174\,\dfrac{lb_m\,ft}{lb_f\,s^2} \cdot \dfrac{\text{..............}}{\text{..............}lb_m} \cdot \dfrac{m}{1\,ft} \cdot \dfrac{\text{..............}}{4.4482\,N}$

$= 1\,\dfrac{kg\,m}{N\,s^2}$

4) $1000\dfrac{lb_m\,ft}{s^2} = 1000\,\dfrac{lb_m\,ft}{s^2} \cdot \dfrac{0.4536\,kg}{\text{..............}} \cdot \dfrac{\text{..............}}{1\,ft} \cdot \dfrac{1\,N}{1\,kg\,m/s^2} = 138.3\,N$

5) $10\dfrac{kcal}{min\,ft\,°F} = 10\,\dfrac{kcal}{min\,ft\,°F} \cdot \dfrac{1055.06\,J}{0.252\,kcal} \cdot \dfrac{\text{.........}min}{60\,s} \cdot \dfrac{\text{...........}ft}{\text{..........}m} \cdot \dfrac{\text{..........}°F}{\text{.........}K}$

$\cdot \dfrac{\text{.........}W}{\text{.........}J/s} = 4121\,\dfrac{W}{m\,K}$

6) $30\,psia = 30\,\dfrac{lb_f}{in^2} \cdot \dfrac{\text{.................}in^2}{\text{................}m^2} \cdot \dfrac{\text{.................}N}{\text{................}lb_f} \cdot \dfrac{\text{................}Pa}{\text{............}N/m^2}$

$\cdot \dfrac{\text{................}atm}{\text{.............}Pa} = 2.04\,atm$

7) $0.002\dfrac{kg}{m\,s} = 0.002\,\dfrac{kg}{m\,s} \cdot \dfrac{\text{.............}lb_m}{\text{.............}kg} \cdot \dfrac{\text{.............}m}{\text{...............}ft} = 0.0013\dfrac{lb_m}{ft\,s}$

8) $5\dfrac{lb\,mol}{h\,ft^2\,mol\,frac} = 5\,\dfrac{lb\,mol}{h\,ft^2\,mol\,frac} \cdot \dfrac{\text{.............}kg\,mol}{\text{.............}lb\,mol} \cdot \dfrac{\text{.............}h}{\text{...............}s}$

$\cdot \dfrac{\text{.............}ft^2}{\text{...............}m^2} = 6.78 \times 10^{-3}\dfrac{kg\,mol}{s\,m^2\,mol\,frac}$

9) $1.987\,\dfrac{Btu}{lb\,mol\,°R} = 1.987\,\dfrac{Btu}{lb\,mol\,°R} \cdot \dfrac{\text{.............}cal}{\text{.............}Btu} \cdot \dfrac{\text{.............}lb\,mol}{\text{...............}g\,mol} =$

$\cdot \dfrac{\text{.............}°R}{\text{...........}K} = 1.987\,\dfrac{cal}{g\,mol\,K}$

10) $10.731\,\dfrac{ft^3\,lb_f}{in^2\,lb\,mol\,°R} = 10.731\,\dfrac{ft^3\,lb_f}{in^2\,lb\,mol\,°R} \cdot \dfrac{\text{...............}m^3}{\text{...............}ft^3} \cdot \dfrac{\text{...............}N}{\text{...............}lb_f}$

$\cdot \dfrac{\text{...............}in^2}{\text{...............}m^2} \cdot \dfrac{\text{...........}lb\,mol}{\text{...........}kg\,mol} \cdot \dfrac{1.8°R}{K}$

$= 8314\dfrac{J}{kg\,mol\,K}$

Exercise 1.2

Make the following conversions:

251°F to °C

(Ans. 121.7 °C)

500°R to K

(Ans. 277.6 K)

0.04 lb_m/in^3 to kg/m^3

(Ans. 1107.2 kg/m^3)

12000 Btu/h to W

(Ans. 3516.9 W)

32.174 ft/s^2 to m/s^2

(Ans. 9.807 m/s^2)

0.01 ft^2/h to m^2/s

(Ans. 2.58x10^{-7} m^2/s)

0.8 cal/g°C to J/kgK

(Ans. 3347.3 J/kgK)

20000 kg m/s^2 m^2 to psi

(Ans. 2.9 psi)

0.3 Btu/lb_m°F to J/kgK

(Ans. 1256 J/kgK)

1000 ft^3/(h ft^2 psi/ft) to

cm^3/(s cm^2 Pa/cm)

(Ans. 0.0374 cm^3/(s cm^2 Pa/cm)

Chapter 2
Use of Steam Tables

Review Questions

Which of the following statements are true and which are false?

1. The heat content of liquid water is sensible heat.
2. The enthalpy change accompanying the phase change of liquid water at constant temperature is the latent heat.
3. Saturated steam is at equilibrium with liquid water at the same temperature.
4. Absolute values of enthalpy are known from thermodynamic tables, but for convenience the enthalpy values in steam tables are relative values.
5. The enthalpy of liquid water at 273.16 K in equilibrium with its vapor has been arbitrarily defined as a datum for the calculation of enthalpy values in the steam tables.
6. The latent heat of vaporization of water is higher than the enthalpy of saturated steam.
7. The enthalpy of saturated steam includes the sensible heat of liquid water.
8. The enthalpy of superheated steam includes the sensible heat of vapor.
9. Condensation of superheated steam is possible only after the steam has lost its sensible heat.
10. The latent heat of vaporization of water increases with temperature.
11. The boiling point of water at certain pressure can be determined from steam tables.
12. Specific volume of saturated steam increases with pressure.
13. The enthalpy of liquid water is greatly affected by pressure.
14. The latent heat of vaporization at a certain pressure is equal to the latent heat of condensation at the same pressure.
15. When steam is condensing, it gives off its latent heat of vaporization.
16. The main reason steam is used as a heating medium is its high latent heat value.
17. About 5.4 times more energy is needed to evaporate 1 kg of water at 100 °C than to heat 1 kg of water from 0 °C to 100 °C.
18. The latent heat of vaporization becomes zero at the critical point.
19. Superheated steam is preferred to saturated steam as a heating medium in the food industry.

20. Steam in the food industry is usually produced in "water in tube" boilers.
21. Water boils at 0°C when the absolute pressure is 611.3 Pa
22. Water boils at 100°C when the absolute pressure is 101325 Pa.
23. Steam quality expresses the fraction or percentage of vapor phase to liquid phase of a vapor-liquid mixture.
24. A Steam quality of 70% means that 70% of the vapor-liquid mixture is in the liquid phase (liquid droplets) and 30% in the vapor phase.
25. The quality of superheated steam is always 100%.

Examples

Example 2.1

From the steam tables:

> Find the enthalpy of liquid water at 50 °C, 100 °C, and 120 °C.
> Find the enthalpy of saturated steam at 50 °C, 100 °C, and 120 °C.
> Find the latent heat of vaporization at 50 °C, 100 °C, and 120 °C.

Solution

Step 1
From the column of the steam tables that gives the enthalpy of liquid water read:

> Hat50°C = 209.33kJ/kg
> Hat100°C = 419.04kJ/kg
> Hat120°C = 503.71kJ/kg

Step 2
From the column of the steam tables that gives the enthalpy of saturated steam read:

> Hat50°C = 2592.1kJ/kg
> Hat100°C = 2676.1kJ/kg
> Hat120°C = 2706.3kJ/kg

Step 3
Calculate the latent heat of vaporization as the difference between the enthalpy of saturated steam and the enthapy of liquid water.

> Latent heat at 50°C = 2592.1 − 209.33 = 2382.77kJ/kg
> Latent heat at 100°C = 2676.1 − 419.09 = 2257.06kJ/kg
> Latent heat at 120°C = 2706.3 − 503.71 = 2202.59kJ/kg

Example 2.2

Find the enthalpy of superheated steam with pressure 150 kPa and temperature 150 °C.

Solution

Step 1
Find the enthalpy from the steam tables for superheated steam:

$$H_{steam} = 2772.6 kJ/kg$$

Step 2
Alternatively find an approximate value from:

$$H_{steam} = H_{saturated} + c_{p\ vapor}(T - T_{saturation}) = 2693.4 + 1.909 \times (150 - 111.3)$$
$$= 2767.3\,kJ/kg$$

Example 2.3

If the enthalpy of saturated steam at 50 °C and 55 °C is 2592.1 kJ/kg and 2600.9 kJ/kg respectively, find the enthalpy at 53 °C.

Solution

Find the enthalpy at 53 °C by interpolation between the values for 50 °C and 55°C given in steam tables, assuming that the enthalpy in this range changes linearly:

$$H = 2592.1 + \frac{53 - 50}{55 - 50}(2600.9 - 2592.1) = 2597.4\,kJ/kg$$

Exercises

Exercise 2.1

Find the boiling temperature of a juice that is boiling at an absolute pressure of 31.19 Pa. Assume that the boiling point elevation is negligible.

Solution

From the steam tables, find the saturation temperature at water vapor pressure equal to 31.19 kPa as T = ...°C. Therefore the boiling temperature will be ....

Exercise 2.2

A food product is heated by saturated steam at 100 °C. If the condensate exits at 90 °C, how much heat is given off per kg steam?

Solution

Step 1
Find the the enthalpy of steam and condensate from steam tables:

$$H_{steam} = \text{..............................} kJ/kg,$$

$$H_{condensate} = \text{..............................} kJ/kg.$$

Step 2
Calculate the heat given off:

$$H = \text{.....................} - \text{.....................} = 2299.2 \text{ kJ/kg}$$

Exercise 2.3

Find the enthalpy of steam at 169.06 kPa pressure if its quality is 90%.

Solution

Step 1
Find the enthalpy of saturated steam at 169.06 kPa from the steam tables:

$$H_{steam} = \text{.....................................}$$

Step 2
Find the enthalpy of liquid water at the corresponding temperature from the steam tables:

$$H_{liquid} = \text{.......................................}$$

Step 3
Calculate the enthalpy of the given steam:
$$H = x_s H_s + (1 - x_s)H_L =$$
$$= \text{.................} \times \text{.................} + \text{.................} \times \text{.................} = 2477.3 \text{ } kJ/kg$$

Exercise 2.4

Find the vapor pressure of water at 72 °C if the vapor pressure at 70 °C and 75 °C is 31.19 kPa and 38.58 kPa respectively.

(Hint: Use linear interpolation.)

Exercise 2.5

The pressure in an autoclave is 232 kPa, while the temperature in the vapor phase is 120°C. What do you conclude from these values?

Solution

The saturation temperature at the pressure of the autoclave should be .... Since the actual temperature in the autoclave is lower than the saturation temperature at 232 kPa, the partial pressure of water vapor in the autoclave is less than 232 kPa. Therefore air is present in the autoclave.

Exercise 2.6

Lettuce is being cooled by evaporative cooling in a vacuum cooler. If the absolute pressure in the vacuum cooler is 934.9 Pa, determine the final temperature of the lettuce.

(Hint: Find the saturation temperature from steam tables.)

Chapter 3
Mass Balance

Review Questions

Which of the following statements are true and which are false?

1. The mass balance is based on the law of conservation of mass.
2. Mass balance may refer to total mass balance or component mass balance.
3. Control volume is a region in space surrounded by a control surface through which the fluid flows.
4. Only streams that cross the control surface take part in the mass balance.
5. At steady state, mass is accumulated in the control volume.
6. In a component mass balance, the component generation term has the same sign as the output streams.
7. It is helpful to write a mass balance on a component that goes through the process without any change.
8. Generation or depletion terms are included in a component mass balance if the component undergoes chemical reaction.
9. The degrees of freedom of a system is equal to the difference between the number of unknown variables and the number of independent equations.
10. In a properly specified problem of mass balance, the degrees of freedom must not be equal to zero.

Examples

Example 3.1

How much dry sugar must be added in 100 kg of aqueous sugar solution in order to increase its concentration from 20% to 50%?

Solution

Step 1
Draw the process diagram:

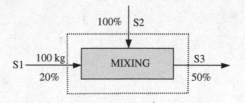

Step 2
State your assumptions:

- dry sugar is composed of 100% sugar.

Step 3
Write the total and component mass balances in the envelope around the process:

i) Overall mass balance

$$100 + S2 = S3 \qquad (3.1)$$

ii) Soluble solids mass balance

$$0.20 \times 100 + S2 = 0.50 \times S3 \qquad (3.2)$$

Solving eqns (3.1) and (3.2) simultaneously, find S2 = 60 kg and S3 = 160 kg. Therefore 60 kg of dry sugar per 100 kg of feed must be added to increase its concentration from 20% to 50%.

Example 3.2

Fresh orange juice with 12% soluble solids content is concentrated to 60% in a multiple effect evaporator. To improve the quality of the final product the concentrated juice is mixed with an amount of fresh juice (cut back) so that the concentration of the mixture is 42%. Calculate how much water per hour must be evaporated in the evaporator, how much fresh juice per hour must be added back and how much final product will be produced if the inlet feed flow rate is 10000 kg/h fresh juice. Assume steady state.

Solution

Step 1
Draw the process diagram:

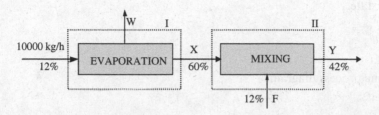

Step 2
Write the total and component mass balances in envelopes I and II:

i) Overall mass balance in envelope I

$$10000 = W + X \tag{3.3}$$

ii) Soluble solids mass balance in envelope I

$$0.12 \times 10000 = 0.60 \times X \tag{3.4}$$

iii) Overall mass balance in envelope II

$$X + F = Y \tag{3.5}$$

iv) Soluble solids mass balance in envelope II

$$0.60 \times X + 0.12 \times F = 0.42 \times Y \tag{3.6}$$

From eqn (3.4) find X = 2000 kg/h. Substituting X in eqn (3.3) and find W = 8000 kg/h. Solve eqns (iii) and (iv) simultaneously and Substitute X in eqn (3.3) and find = 1200 kg/h and Y = 3200 kg/h.
Therefore 8000 kg/h of water will be evaporated, 1200 kg/h of fresh juice will be added back and 3200 kg/h of concentrated orange juice with 42% soluble solids will be produced.

Exercise 3.3

1000 kg/h of a fruit juice with 10% solids is freeze-concentrated to 40% solids. The dilute juice is fed to a freezer where the ice crystals are formed and then the slush is separated in a centrifugal separator into ice crystals and concentrated juice. An amount of 500 kg/h of liquid is recycled from the separator to the freezer. Calculate the amount of ice that is removed in the separator and the amount of concentrated juice produced. Assume steady state.

Solution

Step 1
Draw the process diagram:

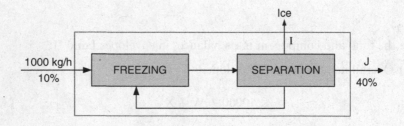

Step 2
Write the total and component mass balances in the envelope around the process:

  i) Overall mass balance

$$1000 = I + J \tag{3.7}$$

  ii) Soluble solids mass balance

$$0.10 \times 1000 = 0.40 \times J \tag{3.8}$$

From eqn (3.8) find $J = 250$ kg/h and then from eqn (3.7) find $I = 750$ kg/h.

Comment: Notice that the recycle stream does not affect the result. Only the streams that cut the envelope take part in the mass balance.

Exercises

Exercise 3.1

How many kg/h of sugar syrup with 10% sugar must be fed to an evaporator to produce 10000 kg/h of sugar syrup with 65% sugar?

Solution

Step 1
Draw the process diagram:

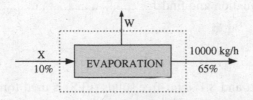

Step 2
State your assumptions:

....

Step 3
Write the mass balance for sugar on the envelope around the process:

$$0.10 \times X = ...$$

Step 4
Solve the above equation and find

$$X = ... \text{ kg/h}$$

Exercise 3.2
How much water must be added to 200 kg of concentrated orange juice with
65% solids to produce orange juice with 12% solids

Solution

Step 1
Draw the process diagram:

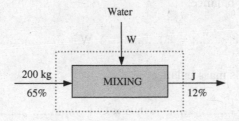

Step 2
Write the mass balance for solids on the envelope around the process:

......................................=..

Solve the above equation and find J = ... kg

Exercise 3.3

Milk with 3.8% fat and 8.1% fat-free solids (FFS) is used for the production of canned concentrated milk. The process includes separation of the cream in a centrifuge and concentration of the partially defatted milk in an evaporator. If the cream that is produced in the centrifuge contains 55% water, 40% fat, and 5% fat-free solids, calculate how much milk is necessary in order to produce a can of concentrated milk that contains 410 g milk with 7.8% fat and 18.1% fat-free solids. How much cream and how much water must be removed in the centrifuge and the evaporator respectively? Assume steady state.

Solution

Step 1
Draw the process diagram:

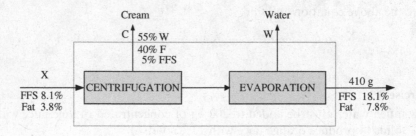

Step 2
Write the total and component mass balances in the envelope around the process:

i) Overall mass balance

.... = ... + W + ... (3.9)

ii) Fat-free solids mass balance

.... = 0.05 × C + ... (3.10)

iii) Fat mass balance

$$0.038 \times X = \text{................................} + \text{..................................} \quad (3.11)$$

Solve eqns (3.9), (3.10) and (3.11) simultaneously and find $X = \text{.................}$ g, $C = \text{...............................}$ g and $W = \text{...............................}$ g.

Exercise 3.4

According to some indications, crystallization of honey is avoided if the ratio of glucose to water is equal to 1.70. Given the composition of two honeys, find the proportions in which they have to be mixed so that the ratio of glucose to water in the blend is 1.7. What will be the composition of the blend?

Honey H_1: glucose 35%, fructose 33%, sucrose 6%, water 16%.
Honey H_2: glucose 27%, fructose 37%, sucrose 7%, water 19%.

Solution

Step 1
Draw the process diagram:

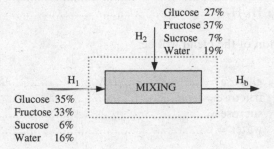

Step 2
Select 1000 kg of blend as a basis for calculation ($H_b = 1000$ kg).

Step 3
Write the total and component mass balances in the envelope around the process:

i) Overall mass balance

$$\text{.............................} + \text{.............................} = \text{.............................} \quad (3.12)$$

ii) Glucose mass balance

$$\text{.............................} + \text{.............................} = \text{.............................} \quad (3.13)$$

iii) Fructose mass balance

$$\dots\dots\dots\dots\dots\dots + \dots\dots\dots\dots\dots = \dots\dots\dots\dots\dots \quad (3.14)$$

iv) Sucrose mass balance

$$\dots\dots\dots\dots\dots\dots + \dots\dots\dots\dots\dots = \dots\dots\dots\dots\dots \quad (3.15)$$

v) Water mass balance

$$\dots\dots\dots\dots\dots\dots + \dots\dots\dots\dots\dots = \dots\dots\dots\dots\dots \quad (3.16)$$

vi) Ratio of glucose to water in the blend

$$G/W = 1.70 \quad (3.17)$$

Solve eqns (3.12) to (3.17) simultaneously and find:

$H_1 = \dots\dots\dots\dots\dots\dots\dots\dots\dots\dots\dots\dots$ kg
$H_2 = \dots\dots\dots\dots\dots\dots\dots\dots\dots\dots\dots\dots$ kg
$H_1/H_2 = \dots\dots\dots\dots\dots\dots\dots\dots\dots\dots\dots\dots$

The composition of the blend will be:

glucose = ...
fructose = ...
sucrose = ...
water = ...

Exercise 3.5

How much glucose syrup with 20% concentration has to be mixed with 100 kg glucose syrup with 40% concentration so that the mixture will have 36% glucose?

Exercise 3.6

How many kg of saturated sugar solution at 70 °C can be prepared from 100 kg of sucrose? If the solution is cooled from 70 °C to 20 °C, how many kg of sugar will be crystallized? Assume that the solubility of sucrose as a function of temperature (in °C) is given by the equation: % sucrose $= 63.2 + 0.146T + 0.0006T^2$.

Exercise 3.7

Find the ratio of milk with 3.8% fat to milk with 0.5% fat that have to be mixed in order to produce a blend with 3.5% fat.

Exercise 3.8

For the production of marmalade, the fruits are mixed with sugar and pectin and the mixture is boiled to about 65% solids concentration. Find the amount of fruits, sugar, and pectin that must be used for the production of 1000 kg marmalade, if the solids content of the fruits is 10%, the ratio of sugar to fruit in the recipe is 56:44, and the ratio of sugar to pectin is 100.

Exercise 3.9

For the production of olive oil, the olives are washed, crushed, malaxated, and separated into oil, water. and solids by centrifugation as in the following flow chart. Find the flow rate in the exit streams given that: a) the composition of the olives is 20% oil, 35% water, and 45% solids; b) the composition of the discharged solids stream in the decanter is 50% solids and 50% water; c) 90% of the oil is taken out in the first disc centrifuge; and d) the ratio of olives to water added in the decanter is equal to 1.

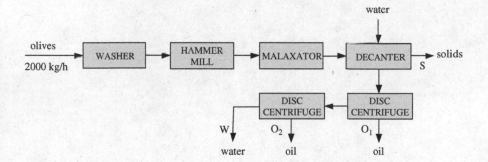

Chapter 4
Energy Balance

Theory

The overall energy balance equation for a system with one inlet (point 1) and one outlet (point 2) is:

$$\left(H_1 + \frac{v_{m1}^2}{2\alpha} + z_1 g \right) \dot{m}_1 - \left(H_2 + \frac{v_{m2}^2}{2\alpha} + z_2 g \right) \dot{m}_2 + q - W_s = \frac{d(mE)}{dt}$$

The overall energy balance equation for a system at steady state with more than two streams can be written as:

$$\sum \left[\left(H + \frac{v_m^2}{2\alpha} + zg \right) \dot{m} \right] = q - W_s$$

where

H	=	enthalpy, J/kg
v_m	=	average velocity, m/s
α	=	correction coefficient (for a circular pipe $\alpha = 1/2$ for laminar flow, $\alpha \approx 1$ for turbulent flow)
z	=	relative height from a reference plane, m
m	=	mass of the system, kg
$\dot{m}$	=	mass flow rate, kg/s
q	=	heat transferred across the boundary to or from the system (positive if heat flows to the system), W
W_s	=	shaft work done by or to the system (positive if work is done by the system), W
E	=	total energy per unit mass of fluid in the system, J/kg
t	=	time, s

In most of the cases, the overall energy balance ends up as an enthalpy balance because the terms of kinetic and potential energy are negligible compared to the enthalpy term, the system is assumed adiabatic ($Q = 0$), and there is no shaft work ($W_s = 0$). Then:

$$\sum \dot{m} H = 0$$

S. Yanniotis, *Solving Problems in Food Engineering*.
© Springer 2008

Review Questions

Which of the following statements are true and which are false?

1. The energy in a system can be categorize as internal energy, potential energy, and kinetic energy.
2. A fluid stream carries internal energy, potential energy, and kinetic energy.
3. A fluid stream entering or exiting a control volume is doing PV work.
4. The internal energy and the PV work of a stream of fluid make up the enthalpy of the stream.
5. Heat and shaft work may be transferred through the control surface to or from the control volume.
6. Heat transferred from the control volume to the surroundings is considered positive by convention.
7. For an adiabatic process, the heat transferred to the system is zero.
8. Shaft work supplied to the system is considered positive by convention.
9. The shaft work supplied by a pump in a system is considered negative.
10. If energy is not accumulated in or depleted from the system, the system is at steady state.

Examples

Example 4.1

1000 kg/h of milk is heated in a heat exchanger from 45°C to 72°C. Water is used as the heating medium. It enters the heat exchanger at 90°C and leaves at 75°C. Calculate the mass flow rate of the heating medium, if the heat losses to the environment are equal to 1 kW. The heat capacity of water is given equal to 4.2 kJ/kg°C and that of milk 3.9 kJ/kg°C.

Solution

Step 1
Draw the process diagram:

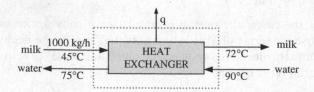

Step 2
State your assumptions:

- The terms of kinetic and potential energy in the energy balance equation are negligible.
- A pump is not included in the system ($W_s = 0$).
- The heat capacity of the liquid streams does not change significantly with temperature.
- The system is at steady state.

Step 3
Write the energy balance equation:

$$Rate\ of\ energy\ input = \dot{m}_{w\ in} \times H_{w\ in} + \dot{m}_{m\ in} \times H_{m\ in}$$

$$Rate\ of\ energy\ output = \dot{m}_{w\ out} \times H_{w\ out} + \dot{m}_{m\ out} \times H_{m\ out} + q$$

(with subscript "w" for water and "m" for milk).

At steady state

$$rate\ of\ energy\ input = rate\ of\ energy\ output$$

or

$$\dot{m}_{w\ in} \times H_{w\ in} + \dot{m}_{m\ in} \times H_{m\ in} = \dot{m}_{w\ out} \times H_{w\ out} + \dot{m}_{m\ out} \times H_{m\ out} + q \quad (4.1)$$

Step 4
Calculate the known terms of eqn (4.1)

i) The enthalpy of the water stream is:

Input: $H_{w\ in} = c_p T = 4.2 \times 90 = 378\ kJ/kg$

Output: $H_{w\ out} = c_p T = 4.2 \times 75 = 315\ kJ/kg$

ii) The enthalpy of the milk stream is:

Input: $H_{m\ in} = c_p T = 3.9 \times 45 = 175.5\ kJ/kg$

Output: $H_{m\ out} = c_p T = 3.9 \times 72 = 280.8\ kJ/kg$

Step 5
Substitute the above values in eqn (4.1), taking into account that:

$$\dot{m}_{w\ in} = \dot{m}_{w\ out} = \dot{m}_w\ and\ \dot{m}_{m\ in} = \dot{m}_{m\ out}$$

$$\dot{m}_w \times 378 + 1000 \times 175.5 = \dot{m}_w \times 315 + 1000 \times 280.8 + 1 \times 3600$$

Step 6
Solve for $\dot{m}_w$

$$\dot{m}_w = 1728.6 \ \ \mathrm{kg/h}$$

Example 4.2

A dilute solution is subjected to flash distillation. The solution is heated in a heat exchanger and then flashes in a vacuum vessel. If heat at a rate of 270000 kJ/h is transferred to the solution in the heat exchanger, calculate: a) the temperature of the solution at the exit of the heat exchanger, and b) the amount of overhead vapor and residual liquid leaving the vacuum vessel. The following data are given: Flow rate and temperature of the solution at the inlet of the heat exchanger is 1000 kg/h and 50°C, heat capacity of the solution is 3.8 kJ/kg°C, and absolute pressure in the vacuum vessel is 70.14 kPa.

Solution

Step 1
Draw the process diagram:

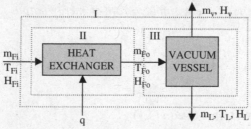

Step 2
State your assumptions:

- The terms of kinetic and potential energy in the energy balance equation are negligible.
- A pump is not included in the system ($W_s = 0$).
- The heat losses to the environment are negligible.
- The heat capacities of the liquid streams do not change significantly with temperature and concentration.
- The system is at steady state.

Step 3
Write the energy balance equation in envelope II:

$$\dot{m}_{Fi}H_{Fi} + q = \dot{m}_{Fo}H_{Fo} \qquad\qquad (4.2)$$

or

$$\dot{m}_{Fi}c_{pF}T_{Fi} + q = \dot{m}_{Fo}c_{pF}T_{Fo} \tag{4.3}$$

Substitute known values:

$$1000 \times 3.8 \times 50 + 270000 = 1000 \times 3.8 \times T_{Fo} \tag{4.4}$$

Solve for T_{Fo}:

$$T_{Fo} = 121\,°C$$

Step 4
Write the mass and energy balance equations in envelope I:

  i) Overall mass balance:

$$\dot{m}_{Fi} = \dot{m}_V + \dot{m}_L \tag{4.5}$$

  ii) Energy balance:

$$\dot{m}_{Fi}H_{Fi} + q = \dot{m}_V H_V + \dot{m}_L H_L \tag{4.6}$$

or

$$\dot{m}_{Fi}c_{pF}T_{Fi} + q = \dot{m}_V H_V + \dot{m}_L c_{pL}T_L \tag{4.7}$$

Step 5
Calculate m_v using equations (4.5), (4.6) and (4.7):

  i) From eqn (4.5):

$$\dot{m}_L = \dot{m}_{Fi} - \dot{m}_V \tag{4.8}$$

  ii) Substitute eqn (4.8) in (4.7):

$$\dot{m}_{Fi}c_{pF}T_{Fi} + q = \dot{m}_V H_V + (\dot{m}_{Fi} - \dot{m}_V)c_{pL}T_L \tag{4.9}$$

  iii) Find the saturation temperature and the enthalpy of saturated vapor at 70.14 kPa from the steam tables:

$$T_L = 90°C$$

$$V = 2660 \text{ kJ/kg}$$

  iv) Substitute numerical values in eqn (4.9):

$$1000 \times 3.8 \times 50 + 270000 = \dot{m}_V \times 2660 + (1000 - \dot{m}_V) \times 3.8 \times 90$$

   v) Solve for $\dot{m}_V$

$$\dot{m}_V = 50.9 \text{ kg/h}$$

Step 6

Alternatively, an energy balance in envelope III can be used instead of envelope I:

   i) Write the energy balance equation:

$$\dot{m}_{Fo}c_{pF}T_{Fo} = \dot{m}_V H_V + \dot{m}_L c_{pL} T_L \qquad (4.10)$$

   ii) Combine eqns (4.5) and (4.10) and substitute numerical values:

$$1000 \times 3.8 \times 121 = \dot{m}_V \times 2660 + (1000 - \dot{m}_V) \times 3.8 \times 90$$

   iii) Solve for $\dot{m}_V$

$$\dot{m}_v = 50.9 \text{ kg/h}$$

Exercises

Exercise 4.1

How much saturated steam with 120.8 kPa pressure is required to heat 1000 g/h of juice from 5°C to 95°C? Assume that the heat capacity of the juice is 4 kJ/kg°C.

Solution

Step 1
Draw the process diagram:

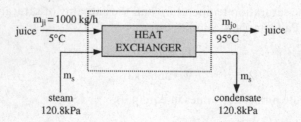

Step 2
Write the energy balance equation:

$$\dot{m}_{ji}H_{ji} + \dot{m}_s H_s = \text{.............................} + \text{.............................}$$

or

$$\dot{m}_{ji}c_{pj}T_{ji} + \dot{m}_s H_s = \text{.............................} + \text{.............................}$$

Step 3
Substitute numerical values in the above equation.
(Find the enthalpy of saturated steam and water [condensate] from steam tables):

$$\text{.............................} + \text{.............................} = \text{.............................} + \text{.............................}$$

Step 4
Solve for $\dot{m}_s$

$$\dot{m}_s = \text{.............................kg/h}$$

Exercise 4.2

How much saturated steam with 120.8 kPa pressure is required to concentrate 1000 kg/h of juice from 12% to 20% solids at 95°C? Assume that the heat capacity of juice is 4 kJ/kg°C.

Solution

Step 1
Draw the process diagram:

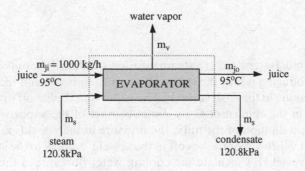

Step 2
Write the overall mass balance equation on the juice side:

$$1000 = \dot{m}_V + \dot{m}_{jo}$$

Step 3
Write the solids mass balance equation:

$$0.12 \times 1000 = \dots\dots\dots \times \dot{m}_{jo}$$

Solve for $\dot{m}_{jo}$ and $\dot{m}_V$

$$\dot{m}_{jo} = \dots\dots\dots\dots\dots\dots\dots\dots\dots\dots\dots\dots\dots\dots\dots.kg/h$$

$$\dot{m}_V = \dots\dots\dots\dots\dots\dots\dots\dots\dots\dots\dots\dots\dots\dots\dots.kg/h$$

Step 4

i) Write the enthalpy balance equation:

$$\dot{m}_{ji}c_{pj}T_{ji} + \dot{m}_s H_s = \dots\dots\dots\dots\dots + \dots\dots\dots\dots\dots = \dots\dots\dots\dots$$

ii) From steam tables, find the enthalpy of water vapor at 95°C, of saturated steam at 120.8 kPa, and of water (condensate) at 120.8 kPa.
iii) Substitute numerical values in the above equation:

$$\dots\dots\dots + \dots\dots\dots = \dots\dots\dots + \dots\dots\dots + \dots\dots\dots$$

iv) Solve for $\dot{m}_s$

$$\dot{m}_s = \dots\dots\dots\dots\dots\dots\dots\dots\dots\dots\dots.kg/h$$

Exercise 4.3

2000 kg/h of milk is sterilized in a steam infusion sterilizer. The milk is heated to 145°C by introducing it into the steam infusion chamber H and then is cooled quickly by flashing in the flash vessel F. The vapor that flashes off in the vessel F is condensed in the condenser C by direct contact of the vapor with cooling water. To avoid dilution of the milk, the pressure in the vessel F must be such that the rate at which vapor flashes off in the vessel F is equal to the steam that is added in the vessel H. Calculate the cooling water flow rate in the condenser that will give the required pressure in the flash vessel. The following data are given: The temperature of the milk at the inlet of H is 40°C, the temperature of the cooling water at the inlet of the condenser is 20°C, the steam introduced into the chamber H is saturated at 475.8 kPa pressure, and the heat capacity of the milk is 3.8 kJ/kg°C at the inlet of the infusion chamber and 4 kJ/kg°C at the exit of the infusion chamber.

Solution

Step 1
Draw the process diagram:

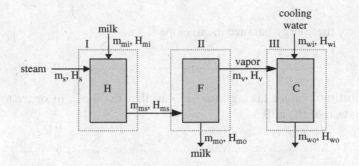

Step 2
State your assumptions:

- The terms of kinetic and potential energy in the energy balance equation are negligible.
- A pump is not included in the system ($W_s = 0$).
- The heat losses to the environment are negligible.
- The water vapor pressure of the milk is equal to that of water at the same temperature.
- The water vapor pressure in the condenser is equal to the water vapor pressure in the flash vessel.
- The system is at steady state.

Step 3
Write the mass and energy balance equations in envelope I:

i) Energy balance in envelope I:

$$\dot{m}_{mi}H_{mi} + \dot{m}_s H_s = \dot{m}_{ms}H_{ms}$$

ii) Overall mass balance in envelope I:

$$\ldots\ldots\ldots\ldots\ldots\ldots + \ldots\ldots\ldots\ldots\ldots = \ldots\ldots\ldots\ldots\ldots$$

iii) Substitute numerical values and combine the last two equations:

$$\ldots\ldots\ldots\ldots\ldots\ldots\ldots\ldots + 2746.5\,\dot{m}_s = \ldots\ldots\ldots\ldots\ldots\ldots\ldots\ldots$$

iv) Solve for $\dot{m}_s$

$$\dot{m}_s = \ldots\ldots\ldots\ldots\ldots\ldots\ldots\ldots\ldots kg/h$$

Step 4

i) Write the energy balance in envelope II:

$$\ldots\ldots\ldots\ldots\ldots\ldots = \ldots\ldots\ldots\ldots\ldots\ldots + \ldots\ldots\ldots\ldots\ldots\ldots$$

ii) Substitute values taking into account that $\dot{m}_s = \dot{m}_v$, in order to avoid dilution of the milk:

$$\ldots\ldots\ldots\ldots\ldots\ldots = \ldots\ldots\ldots\ldots\ldots\ldots + \ldots\ldots\ldots\ldots\ldots\ldots$$

or

$$1389158 - 7600 \times T = 395.1 \times H_V$$

iii) Solve the last equation by trial and error to find the value of T that will give a value of H_V in agreement with steam tables.

$$T = \ldots\ldots\ldots\ldots\ldots\ldots °C.$$

Step 5

Write the overall mass balance and energy balance in envelope III:

i) Overall mass balance:

$$\ldots\ldots\ldots\ldots\ldots\ldots + \ldots\ldots\ldots\ldots\ldots\ldots = \ldots\ldots\ldots\ldots\ldots\ldots$$

ii) Energy balance:

$$\ldots\ldots\ldots\ldots\ldots\ldots + \ldots\ldots\ldots\ldots\ldots\ldots = \ldots\ldots\ldots\ldots\ldots\ldots$$

iii) Substitute numerical values in the last equation and solve for m_{wi}.

The temperature of the water at the exit of the condenser must be equal to $\ldots\ldots\ldots\ldots °C$, because the water vapor pressure in the condenser was assumed equal to that in the flash vessel F.

$$\dot{m}_{wi} = \ldots\ldots\ldots\ldots\ldots\ldots\ldots kg/h$$

Exercise 4.4

Find the amount of saturated steam at 270.1 kPa required to heat 100 kg of cans from 50°C to 121°C, if the heat capacity of the cans is 3.5 kJ/kg°C.

Exercise 4.5

One ice cube at −10°C weighing 30g is added to a glass containing 200ml of water at 20°C. Calculate the final water temperature when the ice cube melts completely. Assume that 3 kJ of heat are transferred from the glass to the water during the melting of the ice? Use the following values: the latent heat of fusion of the ice is 334 kJ/kg, the heat capacity of the ice is 1.93 kJ/kg°C, and the heat capacity of the water is 4.18 kJ/kg°C.

Exercise 4.6

For quick preparation of a cup of hot chocolate in a cafeteria, cocoa powder and sugar are added in a cup of water and the solution is heated by direct steam injection. If the initial temperature of all the ingredients is 15°C, the final temperature is 95°C, the mass of the solution is 150g initially, and the heat capacity of the solution is 3.8 kJ/kg°C, calculate how much saturated steam at 110°C will be used. State your assumptions.

Exercise 4.7

Calculate the maximum temperature to which a liquid food can be preheated by direct steam injection if the initial temperature and the initial solids concentration of the food are 20°C and 33% respectively, and the final solids concentration must not be less than 30%. How much saturated steam at 121 kPa pressure will be used? Assume that the heat capacity of the food is 3.0 kJ/kg°C initially and 3.1 kJ/kg°C after the steam injection.

Chapter 5
Fluid Flow

Review Questions

Which of the following statements are true and which are false?

1. The Reynolds number represents the ratio of the inertia forces to viscous forces.
2. If the Reynolds number in a straight circular pipe is less than 2100, the flow is laminar.
3. The velocity at which the flow changes from laminar to turbulent is called critical velocity.
4. The Reynolds number in non-Newtonian fluids is called the Generalized Reynolds number
5. The velocity profile of Newtonian fluids in laminar flow inside a circular pipe is parabolic.
6. The velocity profile of Newtonian fluids in laminar flow is flatter than in turbulent flow.
7. The maximum velocity of Newtonian fluids in laminar flow inside a circular pipe is twice the bulk average velocity.
8. The average velocity of Newtonian fluids in turbulent flow inside a circular pipe is around 80% of the maximum velocity.
9. The maximum velocity of pseudoplastic fluids in laminar flow inside a circular pipe is more than twice the bulk average velocity.
10. The Hagen-Poiseuille equation gives the pressure drop as a function of the average velocity for turbulent flow in a horizontal pipe.
11. The pressure drop in laminar flow is proportional to the volumetric flow rate.
12. The pressure drop in turbulent flow is approximately proportional to the 7/4 power of the volumetric flow rate.
13. In a fluid flowing in contact with a solid surface, the region close to the solid surface where the fluid velocity is affected by the solid surface is called boundary layer.
14. The velocity gradients and the shear stresses are larger in the region outside the boundary layer than in the boundary layer.
15. Boundary layer thickness is defined as the distance from the solid surface where the velocity reaches 99% of the free stream velocity.

16. The viscosity of a liquid can be calculated if the pressure drop of the liquid flowing in a horizontal pipe in laminar flow is known.
17. The viscosity of non-Newtonian liquids is independent of the shear rate.
18. The flow behavior index in pseudoplastic liquids is less than one.
19. In liquids that follow the power-law equation, the relationship between average velocity and maximum velocity is independent of the flow behavior index.
20. The apparent viscosity of a pseudoplastic liquid flowing in a pipe decreases as the flow rate increases.

Examples

Example 5.1

Saturated steam at 150° C is flowing in a steel pipe of 2 in nominal diameter, schedule No. 80. If the average velocity of the steam is 10 m/s, calculate the mass flow rate of the steam.

Solution

Step 1
Find the inside diameter for a 2 in pipe schedule No. 80 from table:

$$D = 4.925 cm$$

Step 2
Calculate the inside cross-sectional area of the pipe:

$$A = \frac{\pi D^2}{4} = \frac{\pi (0.04925 m)^2}{4} = 0.001905 \ m^2$$

Step 3
Calculate the volumetric flow rate:

$$Q = A \times v_{aver.} = (0.001905 m^2)(10 m/s) = 0.01905 m^3/s$$

Step 4
Find the specific volume of saturated steam at 150 °C from the steam tables:
$$v = 0.3928 \ m^3/kg$$

Step 5
Calculate the mass flow rate:

$$\dot{m} = \frac{Q}{v} = \frac{0.01905 \ m^3/s}{.0.3928 \ m^3/kg} = 0.0485 \ kg/s$$

Example 5.2

A 50% sucrose solution at $20\,^{\circ}C$ is flowing in a pipe with 0.0475 m inside diameter and 10 m length at a rate of 3 m^3/h. Find: a) the mean velocity, b) the maximum velocity, and c) the pressure drop of the sucrose solution. The viscosity and the density of the sucrose solution at $20\,^{\circ}C$ are 15.43 cp and 1232 kg/m^3 respectively.

Solution

Step 1
Calculate the cross-section area of the pipe:

$$A = \frac{\pi D^2}{4} = \frac{\pi (0.0475m)^2}{4} = 1.77 \times 10^{-3} m^2$$

Step 2
Calculate the mean velocity of the liquid:

$$v_m = \frac{Q}{A} = \frac{8.33 \times 10^{-4} m^3/s}{1.77 \times 10^{-3} m^2} = 0.471 \ m/s$$

Step 3
Calculate the Reynolds number:

$$Re = \frac{Dvp}{\mu} = \frac{(0.0475m)(0.471m/s(1232kg/m^3)}{15.43 \times 10^{-3} kg/ms} = 1786$$

Since Re<2100, the flow is laminar and

$$v_{max} = 2v_m = 2 \times 0.471 m/s = 0.942 m/s$$

Step 4
Calculate the pressure drop using the Hagen-Poiseuille equation:

$$\Delta P = \frac{32 v_m \mu L}{D^2} = \frac{32(0.471m/s)(0.01543 Pas)(10m)}{(0.0475m)^2} = 1030. Pa$$

Exercises

Exercise 5.1

Calculate the Reynolds number for water flowing at 5 m^3/h in a tube with 2 in inside diameter if the viscosity and density of water are 1 cp and 0.998 g/ml respectively. At what flow rate does the flow becomes laminar?

Solution

Step 1
Convert the units to SI:

$$Q = 5 \text{ m}^3/\text{h} = \text{.....................................} \text{m}^3/\text{s}$$

$$D = 2 \text{ in} = \text{...................................} \text{m}$$

$$\mu = 1 \text{ cp} = \text{................................} \frac{\text{kg}}{\text{ms}}$$

$$\rho = 0.998 \text{ g/ml} = \text{................................} \text{kg/m}^3$$

Step 2
Calculate the cross-section area of the pipe:

$$A = \text{..} \text{m}^2$$

Step 3
Calculate the mean velocity of the liquid:

$$v_m = \text{....................................} \text{m/s}$$

Step 4
Calculate the Reynolds number:

$$Re = \text{.......................................}$$

Step 5
For the flow to be laminar, Re must be less than or equal to 2100.

i) Calculate the velocity from the Reynolds number:

$$2100 = \frac{(0.0508\text{m})(v_m)(998\text{Kg/m}^3)}{0.001\text{kg/ms}}$$

Solve for v_m:

$$v_m = \text{.......................................} \text{m/s}$$

ii) Calculate the flow rate for the flow to be laminar using v_m found above:

$$Q = Av_m = \text{...............................} \text{m}^3/\text{s or.............} \text{m}^3/\text{h}$$

Exercise 5.2

Calculate the Reynolds number for applesauce flowing at $5 \text{ m}^3/\text{h}$ in a tube with 2 in inside diameter if the consistency index is $13 \text{ Pa s}^{0.3}$, the flow behavior index is 0.3, and the density is 1100 kg/m^3.

Solution

Step 1
Convert the units to SI:

$$Q = 5 \text{ m}^3/\text{h} = \dots\dots\dots\dots\dots\dots\dots\dots\dots\dots\text{m}^3/\text{s}$$

$$D = 2 \text{ in} = \dots\dots\dots\dots\dots\dots\dots\dots\dots\dots\text{m}$$

Step 2
Calculate the cross-section area of the pipe:

$$A = \dots\dots\dots\dots\dots\dots\dots\dots\dots\dots\text{m}^2$$

Step 3
Calculate the mean velocity of the liquid:

$$v_m = \dots\dots\dots\dots\dots\dots\dots\dots\dots\dots\text{m/s}$$

Step 4
Calculate the Reynolds number:
Since $n \neq 1$, the fluid is non-Newtonian. The Generalised Reynolds number will be:

$$Re_G = 2^{3-n}\left(\frac{n}{3n+1}\right)^n \frac{D^n v_m^{2-n} \rho}{k} = \dots\dots\dots\dots\dots\dots\dots\dots\dots\dots\dots\dots\dots\dots$$

Exercise 5.3

Olive oil is flowing in a horizontal tube with 0.0475 m inside diameter. Calculate the mean velocity if the pressure drop per meter of pipe is 1000 Pa. The viscosity of olive oil is 80 cp and its density is 919 kg/m^3.

Solution

Step 1
Assume the flow is laminar and calculate the mean velocity using the Hagen-Poiseuille equation:

$$v_m = \frac{D^2 \Delta P}{32 L \mu} = \dots\dots\dots\dots\dots\dots\dots\dots\dots\dots\text{m/s}$$

Step 2
Verify that the flow is laminar:

$$Re = \dots\dots\dots\dots\dots\dots\dots\dots\dots\dots\dots\dots\dots\dots\dots\dots\dots\dots\dots$$

Exercise 5.4

Honey at 1 liter/min is flowing in a capillary-tube viscometer 2cm in diameter and 50 cm long. If the pressure drop is 40 kPa, determine its viscosity.

Solution
The viscosity can be calculated using the Hagen-Poiseuille equation.

Step 1
Find the mean velocity:

$$v_m = \frac{Q}{\dots\dots} = \dots\dots\dots\dots\dots m/s$$

Step 2
Calculate the viscosity:

$$\mu = \frac{D^2 \Delta P}{32 L v_m} = \dots\dots\dots\dots\dots Pas$$

Exercise 5.5

Tomato concentrate is in laminar flow in a pipe with 0.0475 m inside diameter and 10 m length at a rate of 3 m^3/h. Find: a) the mean velocity, b) the maximum velocity, and c) the pressure drop of the tomato concentrate. The consistency index and the flow behavior index are K = 18.7 Pas$^{0.4}$ and n = 0.4 respectively. Compare the pressure drop for the sucrose solution of Example 5.2 with the pressure drop of tomato concentrate.

Solution

Step 1
Calculate the cross-section area of the pipe:

$$A = \frac{\pi D^2}{4} = \dots\dots\dots\dots m^2$$

Step 2
Calculate the mean velocity of the liquid:

$$v_m = \frac{Q}{A} = \dots\dots\dots\dots\dots m/s$$

Step 3

Use the relationship between mean and maximum velocity for a power-law non-Newtonian fluid in laminar flow to calculate v_{max}:

$$\frac{v_{max}}{v_m} = \frac{3n+1}{n+1}$$

Therefore

$$v_{max} = \dots\dots\dots\dots\dots\dots\dots\dots\dots\dots\dots\dots\dots\dots\dots m/s$$

Step 4

Find the relationship between mean velocity and pressure drop ΔP in laminar flow for non-Newtonian fluids:

$$v_m = \left(\frac{\Delta P}{2kL}\right)^{1/n} \left(\frac{n}{3n+1}\right) R^{\frac{n+1}{n}}$$

where K is consistency index (Pas^n), n is flow behaviour index, L is pipe length (m), and R is pipe diameter (m).

Step 5

Solve for the pressure drop, substitute values, and find ΔP:

$$\Delta P = \dots\dots\dots\dots\dots\dots\dots\dots\dots\dots\dots\dots\dots\dots\dots Pa$$

Step 6

Compare the above pressure drop with the pressure drop calculated for the sucrose solution.

Exercise 5.6

Develop a spreadsheet program to find and plot the velocity distribution as a function of pipe radius for the sucrose solution of Example 5.2 and for the tomato concentrate of Exercise 5.5. Compare the results.

Solution

Step 1

Find the equations for the velocity distribution in laminar flow in a circular pipe

   i) for a Newtonian fluid:

$$v_r = \frac{\Delta P R^2}{4\mu L}\left(1 - \left(\frac{r}{R}\right)^2\right)$$

ii) for a non-Newtonian fluid:

$$v_r = \left(\frac{\Delta P}{2KL}\right)^{1/n} \left(\frac{n}{n+1}\right) R^{(n+1)/n} \left[1 - \left(\frac{r}{R}\right)^{(n+1)/n}\right]$$

Step 2
Calculate the velocity for various values of radius using the above equations.

Step 3
Plot the results.
You must end up with the following figure for the sucrose solution and the tomato concentrate:

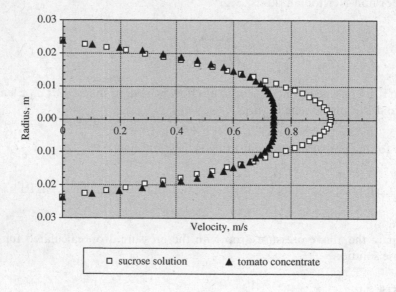

Step 4
Compare the results.

Exercise 5.7

Develop a spreadsheet program to find and plot the velocity distribution as a function of pipe radius for a dilatant liquid with n = 2 and K = 18.7 Pas^2 flowing at a rate of 3 m^3/h in a pipe with 0.0475 m inside diameter and 10 m length.

Chapter 6
Pumps

Theory

The mechanical energy balance equation is used to calculate the required power for a pump. The mechanical energy balance equation for a system with one inlet (point 1) and one outlet (point 2) is:

$$\frac{\left(v_{m2}^2 - v_{m1}^2\right)}{2\alpha} + \frac{P_2 - P_1}{\rho} + (z_2 - z_1)g + \Sigma F = -w_s \qquad (6.1)$$

where v_m = average velocity, m/s
α = kinetic energy correction coefficient (for a circular pipe $\alpha = 1/2$ for laminar flow, $\alpha \approx 1$ for turbulent flow)
P = pressure, Pa
ρ = density, kg/m^3
z = relative height from a reference plane, m
g = acceleration of gravity, 9.81 m/s^2
ΣF = friction losses per unit mass of fluid, J/kg
w_s = work supplied by the pump per unit mass of fluid, J/kg

The available Net Positive Suction Head (NPSH$_a$) is:

$$\text{NPSH}_a = \frac{P - p_v}{\rho g} + z_1 - \frac{\Sigma F_s}{g} \qquad (6.2)$$

where P = pressure in the suction tank, Pa
p_v = vapor pressure of liquid in the pump, Pa
z_1 = distance of the pump from the liquid level in the suction tank, m (z_1 positive if the pump is below the liquid level in the tank, z_1 negative if the pump is above the liquid level in the tank)
ΣF_s = friction losses in the suction line, J/kg

Review Questions

Which of the following statements are true and which are false?

1. Mechanical energy includes kinetic energy, potential energy, shaft work, and the flow work term of enthalpy.
2. Mechanical energy cannot be completely converted to work.
3. The fluid pressure drop due to friction in a straight pipe is proportional to the velocity of the fluid.
4. The pressure drop due to skin friction in a pipe can be calculated from the Fanning equation.
5. The friction factor f in laminar flow depends on the Reynolds number and the surface roughness of the pipe.
6. The friction factor f in turbulent flow can be obtained from the Moody chart.
7. In turbulent flow, the higher the surface roughness of the pipe the higher the influence of the Reynolds number on the friction factor f.
8. A sudden change of the fluid velocity in direction or magnitude causes friction losses.
9. Equation 6.1 gives the energy added to a fluid by a pump.
10. The energy added to a fluid by a pump is often called the developed head of the pump and is expressed in m.
11. The required power for a pump is independent of the liquid flow rate.
12. The brake power of a pump depends on the efficiency of the pump.
13. If the pressure in the suction of a pump becomes equal to the vapor pressure of the liquid, cavitation occurs.
14. Under cavitation conditions, boiling of the liquid takes place in the pump.
15. The difference between the sum of the velocity head and the pressure head in the suction of the pump and the vapor pressure of the liquid is called available net positive suction head (NPSH).
16. To avoid cavitation, the available NPSH must be greater than the required NPSH provided by the pump manufacturer.
17. The higher the temperature of the liquid, the lower the available NPSH.
18. It is impossible to pump a liquid at its boiling point unless the pump is below the liquid level in the suction tank.
19. Centrifugal pumps are usually self-primed pumps.
20. Positive displacement pumps are usually self-primed pumps.
21. Positive displacement pumps develop higher discharge pressures than centrifugal pumps.
22. The discharge line of a positive displacement pump can be closed without damaging the pump.
23. The discharge line of a centrifugal pump can be completely closed without damaging the pump.
24. The flow rate in a positive displacement pump is usually adjusted by varying the speed of the pump.

25. The flow rate in a positive displacement pump decreases significantly as the head increases.
26. Centrifugal pumps are used as metering pumps.
27. Liquid ring pumps are usually used as vacuum pumps.
28. The capacity of a centrifugal pump is proportional to the rotational speed of the impeller.
29. The head developed by a centrifugal pump is proportional to the speed of the impeller.
30. The power consumed by a centrifugal pump is proportional to the cube of the speed of the impeller.

Examples

Example 6.1

A liquid food at 50 °C is being pumped at a rate of 3 m^3/h from a tank A, where the absolute pressure is 12350 Pa, to a tank B, where the absolute pressure is 101325 Pa, through a sanitary pipe 1.5 in nominal diameter with 4.6×10^{-5}m surface roughness . The pump is 1 m below the liquid level in tank A and the discharge in tank B is 3.3 m above the pump. If the length of the pipe in the suction line is 2 m, the discharge line 10 m, and there are one 90 ° elbow in the suction line, two 90 ° elbows in the discharge line, and one globe valve in the discharge line, calculate the power required, the developed head, and the available Net Positive Suction Head (NPSH). Which of the three pumps that have the characteristic curves given below could be used for this pumping job? The viscosity and the density of the liquid are 0.003 mPas and 1033 kg/m^3 respectively. The efficiency of the pump is 65%. Assume that the level in tank A is constant.

Solution

Step 1
Draw the process diagram:

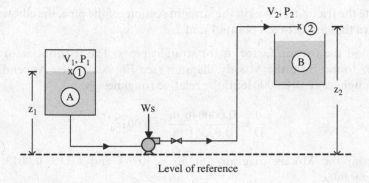

Level of reference

Step 2
Calculate the mean velocity in the pipe:

i) Calculate the mass flow rate, $\dot{m}$:

$$\dot{m} = Q\rho = \left(3\,\frac{m^3}{h}\right)\left(1033\,\frac{kg}{m^3}\right)\left(\frac{1\,h}{3600\,s}\right) = 0.861\ kg/s$$

ii) Find the inside pipe diameter:
The inside pipe diameter of 1.5 in nominal diameter pipe is 1.402 in

$$D = 1.402\,in \times \frac{0.0254\,m}{in} = 0.03561\ m$$

iii) Calculate the cross-section area of the pipe, A:

$$A = \frac{\pi D^2}{4} = \frac{\pi \times (0.03561\,m)^2}{4} = 9.959 \times 10^{-4} m^2$$

iv) Calculate the mean velocity in the pipe, v:

$$v = \frac{Q}{A} = \frac{(3\,m^3/h)/(3600\,s/h)}{(9.959 \times 10^{-4} m^2)} = 0.837\,m/s$$

Step 3
Calculate the Reynolds number:

$$Re = \frac{Dv\rho}{\mu} = \frac{(0.03561\,m)(0.837\,m/s)\left(1033\,kg/m^3\right)}{(0.003\,kg/ms)} = 10263$$

Step 4
Select two points, points 1 and 2, with known v, P, and z values to which to apply the mechanical energy balance equation. The pump must be between points 1 and 2.

Step 5
Calculate the frictional losses in the straight sections of the pipe, the elbows, and the valves that are between points 1 and 2:

i. Find the friction factor, f, for straight pipes. The friction factor f can be found from the Moody diagram (see Fig A.1 in the Appendix). If roughness $\varepsilon = 0.000046$m, the relative roughness is:

$$\frac{\varepsilon}{D} = \frac{0.000046\ m}{0.03561\ m} = 0.0013$$

From the Moody diagram for $Re = 10263$ and $\varepsilon/D = 0.0013$, read $f = 0.008$.

Alternatively, f can be calculated by an empirical relationship e.g., the Colebrook equation:

$$\frac{1}{\sqrt{f}} = -4\log_{10}\left(\frac{\varepsilon/D}{3.7} + \frac{1.255}{Re\sqrt{f}}\right) = -4\log_{10}\left(\frac{0.0013}{3.7} + \frac{1.255}{10263\sqrt{f}}\right)$$

Solving the above equation by trial and error, find $f = 0.0082$.

ii. Find the equivalent length of a 90° standard elbow: $L_e/D = 32$, Equivalent length of straight pipe for 3 elbows:

$$L_e = 3(32D) = 3(32 \times 0.03561) = 3.42 \text{ m}$$

iii. Find the equivalent length of the globe valve: $L_e/D = 300$:

$$L_e = 300D = 300 \times 0.03561 = 10.68 \text{ m}$$

iv. Use the above results to calculate the frictional losses in the straight pipe sections, the elbows, and the valve:

$$h_s = 4f\frac{v^2}{2D}L = 4 \times 0.0082\frac{0.837^2}{2 \times 0.03561}\frac{m^2/s^2}{m}(12 + 3.42 + 10.68)\,m =$$

$$= 8.42\,\frac{m^2}{s^2} = 8.42\,\frac{J}{kg}$$

Units equivalence:

$$\frac{m^2}{s^2} = \frac{m^2 N}{s^2 N} = \frac{m\,mN}{s^2 N} = \frac{mJ}{s^2 N} = \frac{mJ}{s^2 kgm/s^2} = \frac{J}{kg}$$

Step 6
Calculate the frictional losses in the sudden contraction (entrance from the tank to the pipeline) from:

$$h_c = 0.55\left(1 - \frac{A_2}{A_1}\right)^2\frac{v_2^2}{2\alpha}$$

$A_2/A_1 \cong 0$ since $A_1 \gg A_2$. Also, $\alpha = 1$ because the flow is turbulent. Therefore,

$$h_c = 0.55\left(1 - \frac{A_2}{A_1}\right)^2\frac{v_2^2}{2\alpha} = 0.55\frac{1 \times 0.837^2}{2 \times 1}\frac{m^2}{s^2} = 0.19\frac{m^2}{s^2} = 0.19\frac{J}{kg}$$

Step 7
Calculate the total frictional losses:

$$\Sigma F = h_s + h_c = 8.42 + 0.19 = 8.16 J/kg$$

Step 8
Apply the Mechanical Energy Balance Equation between points 1 and 2 in the
diagram. Since the liquid level in the tank is constant, $v_1 = 0$:

$$-w_s = \frac{v_2^2 - v_1^2}{2\alpha} + \frac{P_2 - P_1}{\rho} + (z_2 - z_1)\,g + \Sigma F =$$

$$= \frac{0.837^2 - 0\,\mathrm{m}^2}{2 \times 1\ \ \ \mathrm{s}^2} + \frac{101325 - 12350}{1033}\ \frac{\mathrm{Pa}}{\mathrm{kg/m}^3}$$

$$+ (3.3 - 1)\mathrm{m} \times 9.81\,\frac{\mathrm{m}}{\mathrm{s}^2} + 8.61\,\frac{\mathrm{J}}{\mathrm{kg}} =$$

$$= 117.7\,\frac{\mathrm{J}}{\mathrm{kg}}$$

Units equivalence:

$$\frac{\mathrm{Pa}}{\mathrm{kg/m}^3} = \frac{\mathrm{Pa\ m}^3}{\mathrm{kg}} = \frac{\mathrm{N\ m}^3}{\mathrm{m}^2\,\mathrm{kg}} = \frac{\mathrm{Nm}}{\mathrm{kg}} = \frac{\mathrm{J}}{\mathrm{kg}}$$

Step 9
Calculate the required power:

$$W = -w_s \times \dot{m} = 117.7\,\frac{\mathrm{J}}{\mathrm{kg}} \times 0.861\,\frac{\mathrm{kg}}{\mathrm{s}} = 101.33\,\frac{\mathrm{J}}{\mathrm{s}} = 101.33\ \mathrm{W}$$

Since the efficiency of the pump is 65%, the required power (brake power)
will be:

$$W_a = \frac{W}{\eta} = \frac{101.33}{0.65} = 155.9\ \mathrm{W}$$

Step 10
Calculate the developed head H_m:

$$H_m = \frac{-w_s}{g} = \frac{117.7\ \mathrm{J/kg}}{9.81\ \mathrm{m/s}^2} = 12.0\,\frac{\mathrm{J}}{\mathrm{N}} = 12.0\,\frac{\mathrm{Nm}}{\mathrm{N}} = 12.0\,\mathrm{m}$$

Step 11
Calculate the available Net Positive Suction Head ($NPSH_a$) using eqn (6.2):

i. The total pressure in the suction tank is $P = 12350$ Pa.
ii. The vapor pressure of liquid in the suction is: $p_v = 12349$ Pa (assumption: the vapor pressure of the liquid food at 50 °C is the same as that of
pure water at the same temperature).

iii. The frictional losses in the suction line are:

a) Frictional losses in the straight pipe section and the elbow of the suction line:
The straight pipe section of the suction line is 2 m.
The equivalent straight pipe length of one 90° standard elbow for $L_e/D = 32$, as found in step 5, is

$$L_e = 1(32D) = 32 \times 0.03561 = 1.14 \text{ m}$$

Therefore,

$$h_{ss} = 4f\frac{v^2}{2D}L = 4 \times 0.0082\frac{0.837^2}{2 \times 0.03561}(2 + 1.14) = 1.01\frac{J}{kg}$$

b) Frictional losses in the entrance from the tank to the pipeline:

$$h_c = 0.19 J/kg \text{ (as calculated in step 6)}$$

c) Total losses in the suction line:

$$\Sigma F_s = h_{ss} + h_c = 1.01 + 0.19 = 1.2 \text{ J/kg}$$

iv) Substitute values in eqn (6.2) and calculate $NPSH_a$:

$$NPSH_a = \frac{12350 - 12349}{1033 \times 9.81} + 1 - \frac{1.20}{9.81} = 0.88 \text{ m}$$

Step 12
Select the pump:

Find the volumetric flow rate for each one of the pumps A, B, and C for a developed head of 12 m. Find also the required NPSH at the corresponding flow rate:

- Pump A: gives 1.6 m³/h and requires 0.20 m NPSH. Therefore, it does not give the required flow rate of 3 m³/h when the developed head is 12 m.
- Pump B: gives 3.1 m³/h and requires 1.05 m NPSH. Therefore, it gives the required flow rate of 3 m³/h, but requires more NPSH than the available of 0.81 m. If used, it will cavitate.
- Pump C: gives 3.2 m³/h and requires 0.45 m NPSH. Therefore, it gives the required flow rate of 3 m³/h and requires less NPSH than the available of 0.81 m. Therefore, Pump C is suitable for this pumping job.

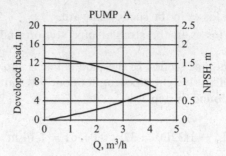

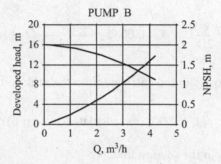

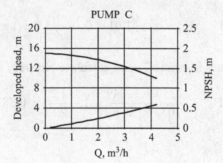

Exercises

Exercise 6.1

Water at 20 °C is flowing in a horizontal pipe 10 m long with 2 in inside diameter. Calculate the pressure drop in the pipe due to friction for a flow rate of 10 m³/h.

Solution

Step 1
Calculate the mean velocity in the pipe:

i) Calculate the cross section area of the pipe:

$$A = \text{..} m^2$$

ii) Calculate the mean velocity in the pipe, v:

$$v = \text{...} m/s$$

Step 2
Calculate the Reynolds number (find density and viscosity of water from a table with physical properties of water):

$$Re = \text{..}$$

Step 3
Calculate the pressure drop from the Fanning equation (since the flow is turbulent):

i) Find the friction factor f from the Moody diagram or from Colebrook equation:

$$f = \text{................................}$$

ii) Calculate the pressure drop:

$$\Delta P = 4f\rho \frac{L}{D} \frac{v^2}{2} = \text{...}Pa$$

Exercise 6.2

You have available a 550 W pump with 70% efficiency. Is it possible to use this pump to transfer $10\ m^3/h$ of a liquid through a 4.7 cm inside diameter pipe, from one open tank to another, if the liquid is discharged at a point 10 m above the liquid level in the suction tank and the total friction losses are 50 J/kg? The density and the viscosity of the liquid are $1050\ kg/m^3$ and 2 cp respectively.

Solution

Step 1
Draw the process diagram.

Step 2
State your assumptions.
....

Step 3
Select points 1 and 2.

Step 4
Calculate the Reynolds number.

i) Calculate the cross section area of the pipe, A:

$$A = \text{...}m^2$$

ii) Calculate the mean velocity in the pipe, v:

$$v = \text{...} \, m/s$$

iii) Calculate the Reynolds number:

$$Re = \text{...}$$

Since the flow is turbulent, $\alpha = $ ....

Step 5
Apply the Mechanical Energy Balance Equation between points 1 and 2.

$$-w_s = \text{...}$$

$$= \text{...}$$

$$\text{...} = 149 \, \frac{J}{kg}$$

Step 6
Calculate the power.

$$- W_s = -w_s \times \dot{m} = \text{...} \, W$$

For a 70% pump efficiency, the required power (brake power) will be:
$$W_{sR} = \frac{\text{..................}}{\text{..............}} = 622 \, W$$

Since the required power is higher than the available 550 W, the pump is not suitable for this pumping job.

Exercise 6.3

A power-law fluid with consistency index $K = 0.223$ Pa $s^{0.59}$, flow behavior index n $= 0.59$, and density $\rho = 1200$ kg/m^3 is pumped through a sanitary pipe having an inside diameter of 0.0475 m at a rate of 5 m^3/h from a tank A to a tank B. The level of the liquid in tank A is 2 m below the pump, while the discharge point is 4 m above the pump. The suction line is 3 m long with one 90° elbow, while the discharge line is 6 m long with two 90° elbows. Calculate the developed head and the discharge pressure of the pump. It is given that the pump is a self-priming pump.

Solution

Step 1
Draw the process diagram.

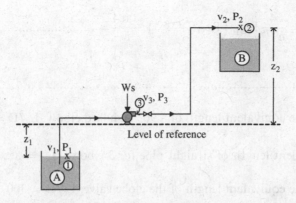

Step 2
Calculate the Reynolds number.

i) Calculate the mass flow rate, $\dot{m}$:

$$\dot{m} = Q\rho = \text{...} \text{ kg/s}$$

ii) Calculate the cross-section area of the pipe, A:

$$A = \text{..} m^2$$

iii) Calculate the mean velocity in the pipe, v :

$$v = \text{..} \text{ m/s}$$

iv) Calculate the Generalized Reynolds number:

$$Re_G = \frac{D^n v^{2-n} \rho}{8^{n-1} \left(\frac{3n+1}{4n}\right)^n K}$$

$$= \frac{(0.0475\,\text{m})^{0.59} (\text{................}\,\text{m/s})^{2-0.59} \left(\text{.................}\,\text{kg/m}^3\right)}{\text{...}} = \text{.........}$$

Step 2
Calculate the frictional losses in the straight pipe sections, the elbows, and the valve.

i) Find the friction factor f.

The friction factor can be calculated from $f = 16/Re_G$ in laminar flow or the empirical relationships of Dodge and Metzner (Ref. 1) in turbulent flow.

$$\frac{1}{\sqrt{f}} = \frac{4}{n^{0.75}} \log_{10}\left(Re_G(f)^{1-n/2}\right) - \frac{0.4}{n^{1.2}} = \dots\dots\dots\dots\dots$$

$$\dots\dots\dots\dots\dots\dots\dots\dots\dots\dots\dots\dots\dots\dots\dots\dots\dots$$

ii) Find the equivalent length of a 90° standard elbow: $L_e/D = 32$

Equivalent length of straight pipe for 3 elbows: $L_e = \dots\dots\dots$ m

iii) Find the equivalent length of the globe valve: $L_e/D = 300$

Equivalent length of straight pipe for 1 globe valve $L_e = \dots\dots$ m

iv) Calculate the frictional losses in the straight pipe sections, the elbows, and the valve:

$h_s = \dots\dots\dots\dots\dots\dots\dots\dots\dots\dots\dots\dots\dots\dots\dots\dots\dots\dots\dots\dots$

$$\dots\dots\dots\dots\dots\dots\dots\dots\dots\dots\dots\dots\dots\dots\dots\dots\dots\dots \frac{J}{kg}$$

Step 3
Calculate the frictional losses in the sudden contraction. Since the flow is laminar, the kinetic energy correction coefficient is:

$$\alpha = \frac{(2n+1)(5n+3)}{3(3n+1)^2} = \dots\dots\dots\dots\dots\dots\dots\dots\dots\dots$$

$$h_c = \dots\dots\dots\dots\dots\dots\dots\dots\dots\dots\dots\dots\dots\dots\dots\dots\dots\dots \frac{J}{kg}$$

Step 4
Calculate the total frictional losses.

$$Ft = \dots\dots\dots\dots\dots\dots\dots\dots\dots\dots\dots\dots\dots\dots\dots\dots\dots\dots\dots \frac{J}{kg}$$

Step 5
Apply the Mechanical Energy Balance Equation between points 1 and 2 in the diagram. Since the liquid level in the tank is constant, $v_1 = 0$.

$$- w_s =$$

$$= \dots\dots\dots\dots\dots\dots\dots\dots\dots\dots\dots\dots\dots\dots\dots\dots\dots =$$

$$= \dots\dots\dots\dots\dots\dots\dots\dots\dots\dots\dots \frac{J}{kg}$$

Step 6
Calculate the required power.

$$- W_s = - w_s \times \dot{m} = \dots\dots\dots\dots\dots\dots\dots\dots\dots\dots W$$

For a 70% pump efficiency, the required power (brake power) will be:

$$W_a = \frac{-W_s}{\eta} = \dots\dots\dots\dots\dots\dots\dots\dots W$$

Step 7
Calculate the developed head H_m.

$$H_m = \frac{-w_s}{g} = \dots\dots\dots\dots\dots\dots\dots\dots\dots\dots m$$

Step 8
Calculate the discharge pressure. Apply the Mechanical Energy Balance Equation (MEBE) between points 2 and 3 with $v_2 = v_3$, $z_3 = 0$, and $- w_s = 0$:

$$\frac{P_3}{\rho} = \frac{P_2}{\rho} + z_2 g + F_d$$

where F_d = the friction losses in the discharge line.

$$P_3 = \dots\dots\dots\dots\dots\dots\dots\dots\dots\dots\dots\dots\dots\dots =$$

$$= \dots\dots\dots\dots Pa$$

Exercise 6.4

Study the spreadsheet program given in *Pump.xls* to get familiar with the way the program works. Modify the spreadsheet program given in *Pump.xls* to solve Example 6.1:

  a) if a heat exchanger, which gives a pressure drop of 50 kPa, is included in the discharge line;

b) if the pump is 1 m above the liquid level in the suction tank (is this pumping possible?); and

c) If the required NPSH is 1 m and the absolute pressure in tank A is 101325 kPa, how many meters below the pump could the suction level be without having cavitation problems? If the pump was pumping water at 21°C from a well, how many meters below the pump could the suction level be without having cavitation problems?

Exercise 6.5

Study the spreadsheet program given in *PumpQ.xls* to get familiar with the way the program works. Run the program and see how the friction losses, the required power, and the developed head increase with increasing flow rate. Modify the spreadsheet so that instead of variable flow rate, it has variable inside pipe diameter with constant flow rate.

Chapter 7
Heat Transfer by Conduction

Theory

As with all transport phenomena, the rate of the transferred quantity is proportional to the driving force and inversely proportional to the resistance. For heat transfer by conduction, the driving force is the temperature difference ΔT and the resistance $R = \Delta x / kA$, where Δx is the wall thickness, k is the thermal conductivity and A is the surface area perpendicular to the direction of transfer. For a cylindrical wall, A is equal to the logarithmic mean surface area, while for a spherical wall, A is equal to the geometric mean surface area. Thus:

Rate of heat transfer

For a single wall	$q = \dfrac{\Delta T}{R}$
For a composite wall	$q = \dfrac{\Delta T}{\sum R}$

Resistance to heat transfer

	Resistance, R	Surface area, A
Plane wall	$\Delta x / kA$	A
Cylindrical wall	$\Delta r / kA_{LM}$	$A_{LM} = \dfrac{A_1 - A_2}{\ln A_1 / A_2}$
Spherical wall	$\Delta r / kA_G$	$A_G = \sqrt{A_1 A_2}$

Review Questions

Which of the following statements are true and which are false?

1. Heat is conducted in solids, liquids, and gases by the transfer of the energy of motion from one more energetic molecule to an adjacent less energetic one.

2. Fourier's law is the basic relationship for heat transfer by conduction.
3. Resistance to heat transfer is proportional to thermal conductivity.
4. Air has low thermal conductivity.
5. Metals have higher thermal conductivity than non-metals.
6. Ice has a thermal conductivity much higher than water.
7. The thermal conductivity of gases is higher than the thermal conductivity of solids.
8. Thermal conductivity is a weak function of temperature.
9. In all cases, thermal conductivity varies with temperature gradient.
10. At steady state, the rate of heat transfer is always zero.
11. At steady state, the temperature at various points in a system does not change with time
12. At steady state, the temperature at various points in a system may change with position.
13. The temperature gradient is positive.
14. For the same heat transfer rate, the slope of the temperature gradientin insulating materials is smaller than in non-insulating materials.
15. The temperature distribution in a plane wall varies linearly with distance in the wall if there is no heat generation in the wall and the thermal conductivity is constant.
16. The temperature distribution in a cylindrical wall varies logarithmically with the distance in the wall if there is no heat generation in the wall and the thermal conductivity is constant.
17. The arithmetic mean area differs from the logarithmic mean area by less than 1.4% if $A_2/A_1 < 1.5$.
18. In a composite wall at steady state, the heat transfer rate in each layer depends on the thermal conductivity of the layer.
19. The temperature drop in a plane wall is inversely proportional to the resistance.
20. The slope of the temperature gradient in each layer of a composite plane wall depends on the thermal conductivity of the layer.

Examples

Example 7.1

Calculate the rate of heat transfer through a glass window with 3 m^2 surface area and 5 mm thickness if the temperature on the two sides of the glass is 14 °C and 15 °C respectively and the thermal conductivity of the glass is 0.7 W/m °C. The system is at steady state.

Solution

Step 1
Draw the process diagram:

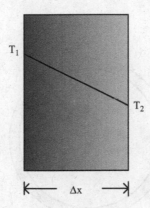

Step 2
Calculate the resistance of the glass to heat transfer:

$$R = \frac{\Delta x}{kA} \doteq \frac{0.005 \text{ m}}{(0.7 \text{ W/m°C})(3 \text{ m}^2)} = 0.00238 \text{ °C/W}$$

Step 3
Calculate the rate of heat transferred:

$$q = \frac{T_1 - T_2}{R} = \frac{15 - 14 \text{ °C}}{0.00238 \text{ °C/W}} = 420 \text{ W}$$

Example 7.2

Hot water is transferred through a stainless steel pipe of 0.04 m inside diameter and 5 m length. The inside wall temperature is 90 °C, the outside surface temperature is 88 °C, the thermal conductivity of stainless steel is 16 W/m °C, and the wall thickness is 2 mm. Calculate the heat losses if the system is at steady state.

Solution

Step 1
Draw the process diagram:

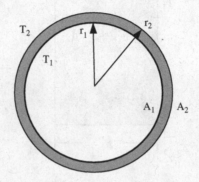

Step 2
Calculate the logarithmic mean area of the wall:

   i) $A_1 = 2\pi\, r_1 L = 2\pi(0.02\,\text{m})(5\,\text{m}) = 0.6283\ \text{m}^2$

   ii) $A_2 = 2\pi\, r_2 L = 2\pi(0.022\,\text{m})(5\,\text{m}) = 0.6912\ \text{m}^2$

   iii) $A_{LM} = \dfrac{A_1 - A_2}{\ln\frac{A_1}{A_2}} = \dfrac{0.6283 - 0.6912}{\ln\frac{0.6283}{0.6912}} = 0.6592\ \text{m}^2$

Step 3
Calculate the resistance of the metal wall to heat transfer:

$$R = \frac{r_2 - r_1}{k_m A_{LM}} = \frac{0.002\,\text{m}}{(16\,\text{W/m}\,^\circ\text{C})(0.6592\text{m}^2)} = 0.00019\,^\circ\text{C/W}$$

Step 4
Calculate the rate of heat transfer:

$$q = \frac{\Delta T}{R} = \frac{T_1 - T_2}{R} = \frac{90 - 88\ ^\circ\text{C}}{0.00019\ ^\circ\text{C/W}} = 10526\ \text{W}$$

Example 7.3

The wall of an oven consists of two metal sheets with insulation in between. The temperature of the inner wall surface is 200 °C and that of the outer surface is 50 °C. The thickness of each metal sheet is 2 mm, the thickness of the insulation is 5 cm, and the thermal conductivity is 16 W/m °C and 0.055 W/m °C respectively. Calculate the total resistance of the wall to heat transfer and the heat transfer losses through the wall per m^2 of wall area.

Solution

Step 1
Draw the process diagram:

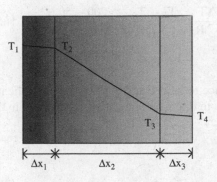

Step 2
State your assumptions.
The system is at steady state.

Step 3
Calculate the resistance to heat transfer:

i) Inner metal wall

$$R_1 = \frac{\Delta x_1}{k_1 A} = \frac{0.002\,m}{(16\,W/m°C)(1\,m^2)} = 0.00013\,°C/W$$

ii) Insulation:

$$R_2 = \frac{\Delta x_2}{k_2 A} = \frac{0.05\,m}{(0.055\,W/m°C)(1\,m^2)} = 0.90909\,°C/W$$

iii) Outer metal wall:

$$R_3 = \frac{\Delta x_3}{k_3 A} = \frac{0.002\,m}{(16\,W/m°C)(1\,m^2)} = 0.00013\,°C/W$$

iv) Total resistance:

$$\Sigma R = R1 + R2 + R3 = 0.00013 + 0.90909 + 0.00013 = 0.90935°C/W$$

Step 4
Calculate the heat transfer through the wall:

$$q = \frac{\Delta T}{\sum R} = \frac{T_1 - T_4}{\sum R} = \frac{200 - 50\,°C}{0.90935\,°C/W} = 165\ W$$

Comments:

1) The main resistance to heat transfer (99.97%) is in the insulation layer.
2) The slope of the temperature gradient is steeper in the layer where the resistance is higher.

Exercises

Exercise 7.1

If an insulation of 2 cm thickness with thermal conductivity equal to 0.02 W/ m °C is wrapped around the pipe of Example 7.2 so that the outside surface temperature of the insulation is 35 °C, while the inside wall temperature is still 90 °C, what would be the heat loss? What will be the outside surface metal wall temperature T_2?

Solution

Step 1
Draw the process diagram:

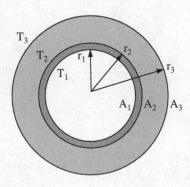

Step 2
Calculate the logarithmic mean area of the insulation:

i) $A_3 =$ .... m^2

ii) $A_{LMi} =$ .... m^2

Step 3
Calculate the resistance of the insulation layer to heat transfer:

$R =$ .... °C/W

Step 4
Calculate the total resistance:

$$\sum R = \text{..}$$

Step 5
Calculate the rate of heat transfer:

$$q = \frac{\Delta T}{\sum R} = \text{..} \ W$$

Step 6
Calculate temperature T_2.
The temperature drop is proportional to the resistance:

$$\frac{T_1 - T_2}{T_1 - T_3} = \frac{R_w}{\sum R}$$

where R_w is the metal wall resistance.

Therefore,

$$T_2 = \text{..} °C$$

Exercise 7.2

The wall of a refrigerator of 4 m^2 surface area consists of two metal sheets with insulation in between. The temperature of the inner wall surface is 5 °C and that of the outer surface is 20 °C. The thermal conductivity of the metal wall is 16 W/m °C and that of the insulation is 0.017 W/m °C. If the thickness of each metal sheet is 2 mm, calculate the thickness of the insulation that is required so that the heat transferred to the refrigerator through the wall is 10 W/m^2.

Solution

Step 1
Draw the process diagram:

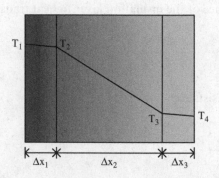

Step 2
State your assumptions

....

Step 3
Calculate the resistance to heat transfer:

    i) Inner metal wall:

$$R_1 = \text{...} \, °C/W$$

    ii) Insulation:

$$R_2 = \text{...} \, °C/W$$

    iii) Outer metal wall:

$$R_3 = \text{...} \, °C/W$$

    iv) Total resistance:

$$\sum R = \text{...} \, °C/W$$

Step 4
Calculate the thickness of insulation:Since

$$\sum R = \frac{\Delta T}{q} = \frac{\text{..}}{\text{..}} \; °C/W$$

the thickness of the insulation will be

$$\Delta x_2 = \text{...} \; m$$

Exercise 7.3

A layer of fat 5 mm thick underneath the skin covers a part of a human body. If the temperature of the inner surface of the fat layer is 36.6 °C and the body loses heat at a rate of 200 W/m², what will be the temperature at the surface of the skin? Assume that the thermal conductivity of fat is 0.2 W/m °C.

Solution

Step 1
State your assumptions:

....

Step 2
Calculate the resistance to heat transfer:

$$R = \text{..} °C/W$$

Step 3
Calculate the surface temperature:
Since

$$q = \frac{\Delta T}{R}$$

the surface temperature of the skin will be

$$T_2 = \text{...} °C$$

Exercise 7.4

A composite plane wall consists of two layers A and B. The thermal conductivity of layers A and B are 0.02 W/m °C and 15 W/m °C respectively. If 100 W/m² are transferred through the wall at steady state, calculate the temperature gradient in the two layers.

Solution

Step 1
State your assumptions:

....

Step 2
From Fourier's law:

$$\frac{dT}{dx} = \text{..}$$

Therefore,

$$\left(\frac{dT}{dx}\right)_A = \text{...}°C/m$$

and

$$\left(\frac{dT}{dx}\right)_B = \text{...}°C/m$$

Exercise 7.5
Find an analytical expression to calculate the heat flux in a plane wall if the thermal conductivity varies with temperature according to the equation $k = k_o + aT$.

Solution

Step 1
Use the expression for k in Fourier's law:

$$\frac{q}{A} = -k\frac{dT}{dx} = -(k_o + aT)\frac{dT}{dx}$$

Step 2
Separate the variables and integrate from x_1 to x_2 and T_1 to T_2:

Exercise 7.6

Develop a spreadsheet program to plot the temperature distribution in a plane wall, a cylindrical wall, and a spherical wall of 4 cm thickness if the outer surface of the wall is at 20 °C and the inner surface is at 200 °C. The inner wall radius for the cylinder and the sphere is 0.1 m, the outer wall radius is 0.14 m, and the thermal conductivity of the wall is 0.02 W/m °C. Assume steady state.

Hint:

The relationships that give the temperature variation as a function of distance in a wall are:

i) Plane wall:

$$T = T_1 - \frac{x - x_1}{kA} q = T_1 - \frac{x}{x_2 - x_1} \Delta T$$

ii) Cylindrical wall:

$$T = T_1 - \frac{\ln(r/r_1)}{2\pi kL} q = T_1 - \frac{\ln(r/r_1)}{\ln(r_2/r_1)} \Delta T$$

iii) Spherical wall:

$$T = T_1 - \frac{\frac{1}{r_1} - \frac{1}{r}}{4\pi k} q = T_1 - \frac{1 - \frac{r_1}{r}}{1 - \frac{r_1}{r_2}} \Delta T$$

Exercise 7.7

Develop a spreadsheet program to plot the temperature distribution in a plane wall of 4 cm thickness and inner wall surface temperature of 200 °C, at steady state, when a) the thermal conductivity of the wall in W/m °C varies according to the equation $k = 0.0325 + 0.0004T$ and 250 W/m^2 of heat are transferred through the wall; and b) the thermal conductivity is constant and equal to the average value for the temperature range of the wall calculated in case a. Calculate the heat transfer rate in case b and compare it to the heat transfer rate of case a.

Hint:

Calculate the temperature as a function of distance in the wall using the expression developed in Exercise 7.5.

Chapter 8
Heat Transfer by Convection

Theory

For heat transfer by convection, the driving force is the temperature difference ΔT, and the resistance R is equal to $1/hA$, where h is the heat transfer coefficient $(W/m^2\,^\circ C)$ and A is the surface area (m^2) perpendicular to the direction of transfer.

Heat transfer rate equation	Resistance, R
Convection only $$q = \frac{\Delta T}{R}$$	$1/hA$
Convection combined with conduction $$q = \frac{\Delta T}{\sum R}$$	Plane wall $$\sum R = \frac{1}{h_i A} + \sum_{j=1}^{n} \frac{\Delta x_j}{k_j A} + \frac{1}{h_o A}$$ Cylindrical wall $$\sum R = \frac{1}{h_i A_i} + \sum_{j=1}^{n} \frac{\Delta r_j}{k_j A_{LM}} + \frac{1}{h_o A_o}$$

where

q = heat transfer rate, W
R = resistance to heat transfer, $^\circ C\,/W$
ΔT = temperature difference, $^\circ C$
h = heat transfer coefficient, $W/m^2\,^\circ C$
h_i = inside surface heat transfer coefficient, $W/m^2\,^\circ C$
h_o = outside surface heat transfer coefficient, $W/m^2\,^\circ C$
A = heat transfer surface area, m^2
A_i = inside heat transfer surface area, m^2
A_o = outside heat transfer surface area, m^2
A_{LM} = logarithmic mean of A_i and A_o , m^2
Δx_j and Δr_j = thickness of layer j, m
k_j = thermal conductivity of layer j, $W/m\,^\circ C$

The heat transfer coefficient is calculated from relationships of the form:

$$Nu = f(Re, Pr) \text{ or } Nu = f(Gr, Pr)$$

where

Nu = Nusselt number
Re = Reynolds number
Gr = Grashof number
Pr = Prandtl number

To calculate the heat transfer coefficient:

1. Determine if the flow is natural or forced (free or forced convection).
2. Identify the geometry of the system.
3. Determine if the flow is laminar or turbulent (calculate the Reynolds number).
4. Select the appropriate relationship $Nu = f(Re, Pr)$
5. Calculate Nu and solve for h.

Review Questions

Which of the following statements are true and which are false ?

1. The rate of heat transfer by convection is calculated using Newton's law of cooling.
2. The heat transfer coefficient depends on the physical properties of the fluid, the flow regime, and the geometry of the system.
3. The units of the heat transfer coefficient are W/m °C.
4. The overall heat transfer coefficient has the same units as the local heat transfer coefficient.
5. The resistance to heat transfer by convection is proportional to the heat transfer coefficient.
6. The heat transfer coefficient in gases is usually higher than in liquids.
7. The heat transfer coefficient is lower in viscous fluids than in water.
8. The heat transfer coefficient in forced convection is higher than in natural convection.
9. The heat transfer coefficient in nucleate boiling is higher than in film boiling.
10. The heat transfer coefficient in dropwise condensation is higher than in film condensation.
11. The movement of a fluid in natural convection results from the differences in the density of the fluid.
12. Fouling increases the overall heat transfer coefficient.

13. Liquid velocities higher than 1 m/s are usually used to reduce fouling.
14. A thermal boundary layer develops on a fluid flowing on a solid surface when the temperature of the fluid is different from the temperature of the solid surface.
15. Temperature gradients exist in the thermal boundary layer.
16. The Prandtl number represents the ratio of thermal diffusivity to momentum diffusivity.
17. The Prandtl number relates the thickness of the hydrodynamic boundary layer to the thickness of the thermal boundary layer.
18. The Grashof number represents the ratio of buoyancy forces to viscous forces.
19. The Grashof number in natural convection plays the role of the Reynolds number in forced convection.
20. The fluid properties at the film temperature are used in calculating the heat transfer coefficient outside various geometries.
21. In heat exchangers, counterflow gives a lower driving force than parallel flow.
22. The logarithmic mean temperature difference is used in heat exchangers as the driving force for heat transfer.
23. The arithmetic mean of ΔT_1 and ΔT_2 differs from their logarithmic mean by more than 1.4% if $\Delta T_1/\Delta T_2 < 1.5$.
24. In the case of multiple-pass heat exchangers, the logarithmic mean temperature difference must be multiplied by a correction factor.
25. In a plate heat exchanger, the surfaces of the plates have special patterns to increase turbulence.
26. Plate heat exchangers are suitable for viscous fluids.
27. Because the distance between the plates in a plate heat exchanger is small, liquids containing particulates may clog the heat exchanger.
28. A shell and tube heat exchanger cannot be used in high pressure applications.
29. Scraped-surface heat exchangers can handle viscous fluids.
30. Fins are used on the outside surface of a heat exchanger pipe when the heat transfer coefficient on the outside surface of the pipe is higher than the heat transfer coefficient inside the pipe.

Examples

Example 8.1

Water flows in a pipe of 0.0475 m inside diameter at a velocity of 1.5 m/s. Calculate the heat transfer coefficient if the temperature of the water is 60 °C and 40 °C at the inlet and the outlet of the pipe respectively, and the inside wall temperature of the pipe is 35 °C.

Solution

Step 1
Draw the process diagram:

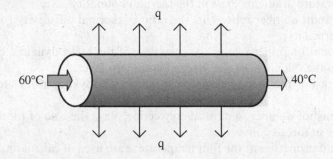

Step 2
Find the physical properties of the water.

The physical properties must be calculated at the average water temperature:

$$T_m = \frac{60 + 40}{2} = 50\,°C$$

Thus,

$\rho_{50} = 988 kg/m^3$
$\mu_{50} = 0.549 cp$
$\mu_{35} = 0.723 cp$
$cp_{50} = 4183 J/kg°C$
$k_{50} = 0.639 W/m°C$

Step 3
Calculate the Reynolds number:

$$Re = \frac{Dv\rho}{\mu} = \frac{(0.0475\,m)(1.5\,m/s)(988\,kg/m^3)}{0.000549\,kg/ms} = 128224$$

Step 4
Identify the regime of heat transfer:

- forced convection
- flow inside a cylindrical pipe
- turbulent flow

Step 5
Select the most suitable equation of Nu = f(Re, Pr) :

$$Nu = 0.023 Re^{0.8} Pr^{0.33} \left(\frac{\mu}{\mu_w}\right)^{0.14}$$

Step 6
Calculate the Prandtl number:

$$Pr = \frac{c_p \mu}{k} = \frac{(4183 \, J/kg\,°C)(0.000549 \, kg/ms)}{0.639 \, W/m\,°C} = 3.59$$

Step 7
Substitute the values of the Reynolds and Prandtl numbers and calculate the Nusselt number:

$$Nu = \frac{hD}{k} = 0.023 (128224^{0.8})(3.59^{0.33})\left(\frac{0.549}{0.723}\right)^{0.14} = 411.7$$

Step 8
Calculate h:

$$h = Nu\frac{k}{D} = 411.7 \frac{0.639 \, W/m\,°C}{0.0475 \, m} = 5538 \, W/m^2\,°C$$

Example 8.2

Sucrose syrup flows in a pipe of 0.023 m inside diameter at a rate of 40 lt/min, while steam is condensing on the outside surface of the pipe. The syrup is heated from 50 to 70 °C, while the inside wall temperature is at 80 °C. Calculate 1) the heat transfer coefficient and 2) the required length of the pipe.

Solution

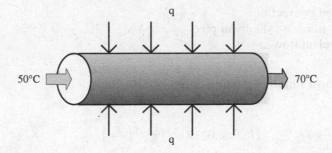

1) Calculation of the heat transfer coefficient:

Step 1
Find the physical properties of sucrose syrup at the average syrup temperature:

$$T_m = \frac{50 + 70}{2} = 60\,°C$$

Thus,

$$\rho_{60} = 1200\,kg/m^3$$

$$\mu_{60} = 3.8\,cp$$

$$\mu_{80} = 2.3\,cp$$

$$c_{p60} = 3120\,J/kg\,°C$$

$$k_{60} = 0.46\,W/m\,°C$$

Step 2
Calculate the Reynolds number:

The mean velocity is:

$$v = \frac{Q}{A} = \frac{(40\,lt/min)(10^{-3}m^3/lt)(1\,min/60\,s)}{\pi(0.023^2\,m^2)/4} = \frac{0.000667\,m^3/s}{0.000415\,m^2} = 1.607\ m/s$$

Therefore,

$$Re = \frac{Dv\rho}{\mu} = \frac{(0.023\,m)(1.607m/s)\left(1200\,kg/m^3\right)}{0.0038\,kg/ms} = 11672$$

Step 3
Identify the regime of heat transfer:

- forced convection
- flow inside a cylindrical pipe
- turbulent flow

Step 4
Select the most suitable equation of $Nu = f(Re, Pr)$:

$$Nu = 0.023Re^{0.8}Pr^{0.33}\left(\frac{\mu}{\mu_w}\right)^{0.14}$$

Step 5
Calculate the Prandtl number:

$$Pr = \frac{c_p \mu}{k} = \frac{(3120 \text{ J/kg }^\circ\text{C})(0.0038 \text{ kg/ms})}{0.460 \text{ W/m }^\circ\text{C}} = 25.8$$

Step 6
Substitute the values of the Reynolds and Prandtl numbers and calculate the Nusselt number:

$$Nu = \frac{hD}{k} = 0.023(11672^{0.8})(25.8^{0.33})\left(\frac{3.8}{2.3}\right)^{0.14} = 129.4$$

Step 7
Calculate h:

$$h = Nu\frac{k}{D} = 129.4\frac{0.460 \text{ W/m }^\circ\text{C}}{0.023 \text{ m}} = 2588 \text{ W/m}^2 \,^\circ\text{C}$$

2) Calculation of the required pipe length:

Step 1
Calculate the heat transferred to the liquid using an enthalpy balance.

   i) Write the enthalpy balance:

$$H_{in} + q = H_{out}$$

or

$$\dot{m}c_p T_{in} + q = \dot{m}c_p T_{out}$$

or

$$q = \dot{m}c_p(T_{out} - T_{in}) \tag{8.1}$$

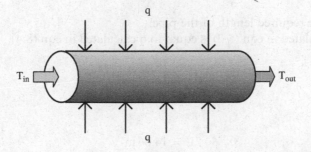

ii) Calculate the mass flow rate ṁ (assume that the density of the syrup at 50°C does not differ significantly from that at 60°C):

$$\dot{m} = Q\rho = \left(40\,\frac{lt}{min}\right)\left(1\,\frac{m^3}{1000\,lt}\right)\left(\frac{1\,min}{60\,s}\right)\left(1200\,\frac{kg}{m^3}\right) = 0.8\,\frac{kg}{s}$$

iii) Substitute values into eqn (8.1) and calculate q:

$$q = \dot{m}c_p(T_{out} - T_{in}) = (0.8\,kg/s)\,(3120\,J/kg\,°C)\,(70 - 50\,°C) = 49920\,W \quad (8.2)$$

Step 2
Calculate the heat transferred to the liquid using the heat transfer rate equation:

i) Write the heat transfer rate equation:

$$q = hA\Delta T_{LM} \quad (8.3)$$

ii) Calculate the driving force for heat transfer (ΔT_{LM}):

$$\Delta T_{LM} = \frac{\Delta T_1 - \Delta T_2}{\ln\frac{\Delta T_1}{\Delta T_2}} = \frac{(80 - 50) - (80 - 70)}{\ln\frac{80-50}{80-70}} = 18.20\,°C$$

iii) Calculate the heat transfer area:

$$A = \pi DL = \pi\,(0.023\,m)\,L = 0.0723L\,m^2$$

iv) Substitute values in eqn (8.3) and calculate q:

$$q = hA\Delta T_{LM} = \left(2588\,W/m^2\,°C\right)(0.0723\,L)(18.2\,°C) = 3405.45\,L \quad (8.4)$$

Step 3
Calculate the required length of the pipe:
Since q calculated in eqn (8.2) is equal to q calculated in eqn (8.4),

$$3405.45\,L = 49920$$

Therefore,

$$L = 14.66\,m$$

Example 8.3

Air is heated by passing over a tube with 0.0127 m outside diameter, while steam is condensing on the inside of the tube. If the heat transfer coefficient on the airside is $15 W/m^2 \,°C$, the overall heat transfer coefficient based on the outside area of the tube is $14.85 W/m^2 \,°C$, the bulk air temperature is $50\,°C$, the steam temperature is $110\,°C$, and the tube surface temperature on the airside is $109.4\,°C$, calculate the heat transferred to the air per m of tube length a) using h_o and b) using U_o.

Solution

Step 1
Calculate the heat transferred based on h_o:

$$q = A_o h_o (T_{tube} - T_{air})$$
$$= (\pi(0.0127\,m)(1\,m))\left(15\,W/m^2\,°C\right)(109.4\,°C - 50\,°C) = 35.5\,W$$

Step 2
Calculate the heat transferred based on U_o:

$$q = A_o U_o (T_{steam} - T_{air})$$
$$= (\pi(0.0127\,m)(1\,m))\left(14.85\,W/m^2\,°C\right)(110\,°C - 50\,°C) = 35.5\,W$$

Comment: Notice that the temperature at the outside surface of the tube is required when the local heat transfer coefficient h is used, while the bulk temperature of the heating medium is required when the overall heat transfer coefficient U is used.

Exercises

Exercise 8.1

In a fruit packaging house, oranges are washed, dried in a stream of high speed air at room temperature, waxed, and dried again in a hot air stream. Calculate the heat transfer coefficient on the surface of an orange if the air velocity is 10 m/s, the air temperature is $55\,°C$, the orange surface temperature is $25\,°C$, and the orange has a spherical shape with a diameter of 8 cm.

Solution

Step 1
Find the physical properties of the air.

The physical properties must be calculated at the film temperature
T_f (average of air temperature and orange surface temperature).

$$T_f = \dots\dots\dots\dots\dots\dots\dots\dots\dots\dots\dots\dots\dots\dots\dots\dots\dots\dots\,°C$$

Thus

$$\rho = \dots\dots\dots\dots\dots\dots\dots\dots kg/m^3$$

$$\mu = \dots\dots\dots\dots\dots\dots\dots Pas$$

$$c_p = \dots\dots\dots\dots\dots\dots\dots J/kg\,°C$$

$$k = \dots\dots\dots\dots\dots\dots\dots W/m°C$$

Step 2
Calculate the Reynolds number:

$$Re = \dots\dots\dots\dots\dots\dots\dots\dots\dots\dots\dots\dots\dots\dots\dots\dots\dots\dots\dots$$

Step 3
Identify the regime of heat transfer:

- $\dots\dots\dots\dots\dots\dots\dots\dots\dots\dots\dots\dots\dots\dots\dots\dots$
- $\dots\dots\dots\dots\dots\dots\dots\dots\dots\dots\dots\dots\dots\dots\dots\dots$
- $\dots\dots\dots\dots\dots\dots\dots\dots\dots\dots\dots\dots\dots\dots\dots\dots$

Step 4
Select the most suitable equation of $Nu = f(Re, Pr)$:

$$Nu = 2 + 0.60Re^{0.50}Pr^{0.33}$$

Step 5
Calculate the Prandtl number:

$$Pr = \dots\dots\dots\dots\dots\dots\dots\dots\dots\dots\dots\dots\dots\dots\dots\dots$$

Step 6
Calculate the Nusselt number:

$$Nu = \dots\dots\dots\dots\dots\dots\dots\dots\dots\dots\dots\dots\dots\dots\dots\dots\dots\dots$$

Step 7
Calculate h:

$$h = \text{..}$$

Exercise 8.2

A horizontal steel pipe with 2 in nominal diameter has an outside surface temperature of 80 °C. Calculate the heat transfer coefficient on the outside surface of the pipe if the pipe is exposed to a room temperature of 20 °C.

Solution

Step 1
Find the physical properties of the air.
The physical properties must be calculated at the film temperature T_f (average of air temperature and pipe surface temperature).

$$T_f = \text{..} \, °C$$

Thus,

$$\rho = \text{............................} kg/m^3$$

$$\mu = \text{............................} Pas$$

$$c_p = \text{............................} J/kg\,°C$$

$$k = \text{............................} W/m\,°C$$

Step 2
Identify the regime of heat transfer:
- ....
- ....

Step 3
Calculate the Grashof number:
Since the pipe is horizontal, the characteristic dimension is the pipe diameter. In a 2 in nominal diameter steel pipe, the outside diameter is 2.375 in or 0.0603 m.

Therefore,

$$Gr = \frac{D^3 \rho^2 g \beta \Delta T}{\mu^2} = \text{..}$$

Step 4
Calculate the Prandtl number:

$$Pr = \text{..}$$

Step 5
Select the most suitable equation of $Nu = f(Gr, Pr)$:

i) Calculate the product:

$$Gr \cdot Pr = \text{..}$$

ii) Select the equation based on the value of $Gr \cdot Pr$ found above:

$$Nu = 0.53Gr^{0.25}Pr^{0.25}$$

Step 6
Calculate the Nusselt number:

$$Nu = \text{...}$$

Step 7
Calculate h:

$$h = \text{...}$$

Exercise 8.3

A home heater consists of a vertical plate with dimensions 0.5m × 1.0m. If the temperature on the surface of the plate is maintained at 70 °C and the room temperature is equal to 21 °C, calculate the heat transferred to the room.

Solution

Step 1
Find the physical properties of the air at the film temperature T_f:

$$T_f = \text{...} °C$$

Thus,

$$\rho = \text{.............................} kg/m^3$$
$$\mu = \text{.............................} Pas$$
$$c_p = \text{.............................} J/kg °C$$
$$k = \text{.............................} W/m °C$$

Step 2
Identify the regime of heat transfer:

- ....
- ....

Step 3
Calculate the Grashof number.
Since the plate is vertical, the characteristic dimension is the plate height.

$$Gr = \text{...}$$

Step 4
Calculate the Prandtl number:

$$Pr = \text{...}$$

Step 5
Select the most suitable equation of $Nu = f(Gr, Pr)$:

i) Calculate the product:

$$Gr \cdot Pr = \text{...}$$

ii) Select the equation suitable for the calculation of the heat transfer coefficient in free convection on a vertical plane with .... $< Gr$ $\cdot Pr <$ ....:

$$Nu = 0.53 Gr^{0.25} Pr^{0.25}$$

Step 6
Calculate the Nusselt number:

$$Nu = \text{...}$$

Step 7
Calculate h:

$$h = \text{..}$$

Step 8
Calculate the heat transfer rate:

$$q = hA\Delta T = \text{...}$$

Exercise 8.4

A liquid food is heated in an unbaffled agitated vessel with a heating jacket. The inside diameter of the vessel is 1 m and the diameter of the flat-bladed paddle agitator is 30 cm. The agitator rotates at $N = 60$ rpm. Calculate the heat transfer coefficient at the wall of the jacket. Assume the following physical properties for the liquid food: density 1030 kg/m^3, viscosity at the bulk liquid temperature 1.5 mPas, viscosity at the wall temperature 1 mPas, heat capacity 4kl/kg °C, thermal conductivity 0.6W/m °C. The following correlation can be used for this case.

$$Nu = 0.36 \, Re^{2/3} Pr^{1/3} \left(\frac{\mu}{\mu_w}\right)^{0.21}$$

Solution

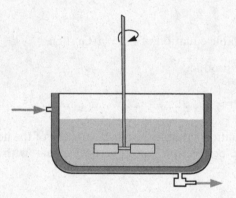

Step 1
Calculate the Re number (take care with the units):

$$Re = \frac{D_{agitator}^2 \, N\rho}{\mu} = \text{..}$$

Step 2
Calculate the Prandtl number:

$$Pr = \frac{\text{...}}{\text{.................................}} =$$

Step 3
Calculate the Nusselt number:

$$Nu = \text{..}$$

Step 4
Calculate the heat transfer coefficient:

$$h = \frac{Nu\ k}{D_{vessel}} = \dots\dots\dots\dots\dots\dots\dots\dots\dots\dots\dots\dots\dots\dots\dots\dots$$

Exercise 8.5

Orange juice is flowing in a pipe with 0.0229 m inside diameter and 0.0254 m outside diameter, while steam is condensing on the outside. If the heat transfer coefficient on the juice side is $1500 W/m^2\,°C$, on the steam side $3000 W/m^2\,°C$ and the thermal conductivity of the tube is $15 W/m\,°C$, calculate the overall heat transfer coefficient based on the outside area U_o and the inside area U_i.

Solution

Step 1
Draw the process diagram:

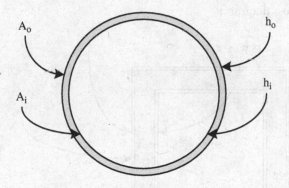

Step 2
Calculate the overall heat transfer coefficient based on the outside surface area of the pipe:

$$U_o = \frac{1}{A_o \sum R} = \frac{1}{A_o\left(\frac{1}{h_i A_i} + \frac{\Delta r}{k A_{LM}} + \frac{1}{h_o A_o}\right)} = \frac{1}{D_o\left(\frac{1}{h_i D_i} + \frac{\ln(D_o/D_i)}{2k} + \frac{1}{h_o D_o}\right)} =$$

$$= \dots\dots\dots\dots\dots\dots\dots\dots\dots\dots\dots\dots\dots\dots\dots\dots\dots\dots\dots\dots$$

Step 3
Calculate the overall heat transfer coefficient based on the inside surface area of the pipe:

$$U_i = \frac{1}{A_i \sum R} = \frac{1}{A_i\left(\frac{1}{h_i A_i} + \frac{\Delta r}{k A_{LM}} + \frac{1}{h_o A_o}\right)} = \frac{1}{D_i\left(\frac{1}{h_i D_i} + \frac{\ln(D_o/D_i)}{2k} + \frac{1}{h_o D_o}\right)} =$$

$$= \dots\dots\dots\dots\dots\dots\dots\dots\dots\dots\dots\dots\dots\dots\dots\dots\dots\dots\dots\dots\dots$$

Exercise 8.6

Air at a temperature of 60 °C flows inside a pilot plant cabinet dryer with dimensions 1m × 1m × 1m at a velocity of 3m/s. The walls of the dryer consist of two metal sheets of 2 mm thickness with 5 cm insulation in between. Calculate the heat losses through the side walls of the dryer if the heat transfer coefficient on the inside surface of the walls is 15W/m² °C, the thermal conductivity of the metal wall and the insulation are 45W/m °C and 0.045W/m °C respectively, and the outside air temperature is 20 °C.

Solution

Step 1
Draw the process diagram:

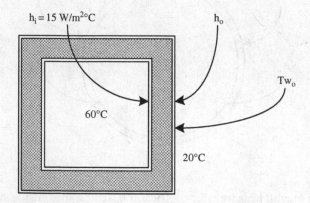

Step 2
Identify the regime of heat transfer:

- forced convection on the inside surface of the wall
- conduction through the wall
- natural convection on the outside surface of the walls

Step 3
Calculate the heat transfer coefficient on the outside surface of the wall:

i) Find the physical properties of the air at the film temperature T_f:
Since the wall temperature is not known, it must be assumed. Assume the outside wall temperature (Tw_o) is 30 °C. This assumption will be verified later (Step 4).

$T_f = $ ....°C

Thus,

$\rho = $ ....kg/m^3

$\mu = $ ....Pas

$c_p = $ ....J/kg°C

$k = $ ....W/m°C

$\beta = $ 1/K

ii) Calculate the Grashof number:
Since the walls are vertical (side walls), the characteristic dimension is

....

Therefore:

$Gr = $ ....

iii) Calculate the Prandtl number:

$Pr = $ ....

iv) Select the most suitable equation of $Nu = f(Gr, Pr)$:
Calculate the product:

$Gr \cdot Pr = $ ....

Since $10^4 < Gr \cdot Pr < 10^9$, the equation suitable for the calculation of the heat transfer coefficient is:

....

v) Calculate the Nusselt number:

$Nu = $ ....

vi) Calculate h:

$$h = \dots\dots\dots\dots\dots\dots\dots\dots\dots\dots\dots\dots\dots\dots\dots\dots\dots\dots\dots$$

Step 4
Verify the assumption made in step 3(i) about the outside wall temperature.

a) Calculate the resistances to heat transfer:

i) Convective resistance on the inside wall surface:

$$R_i = \frac{1}{h_i A} = \dots\dots\dots\dots\dots\dots\dots\dots\dots\dots\dots\dots\dots\dots\dots\dots$$

ii) Conductive resistance of the wall:

$$R_w = \sum_{j=1}^{3} R_j = \frac{\Delta x_m}{k_m A} + \frac{\Delta x_i}{k_i A} + \frac{\Delta x_m}{k_m A} = \dots\dots\dots\dots\dots\dots\dots\dots$$

iii) Convective resistance on the outside wall surface:

$$R_o = \dots\dots\dots\dots\dots\dots\dots\dots\dots\dots\dots\dots\dots\dots\dots$$

iv) Total resistance:

$$\sum R = R_i + R_w + R_o = \dots\dots\dots\dots\dots\dots\dots\dots\dots\dots\dots$$

b) Calculate the temperature drop due to outside convective resistance:

$$\Delta T_o = \frac{R_o}{\sum R} \Delta T_{overall} = \dots\dots\dots\dots\dots\dots\dots\dots\dots\dots$$

c) Calculate the outside wall surface temperature:

$$T_{wo} = T_{air} + \Delta T_o = 20 + 9.6 = 29.6\,°C$$

The calculated T_{wo} with the last relationship is very close to the 30 °C that was assumed in step 3(i). If the T_{wo} calculated here was different from the assumed T_{wo} in step 3(i), another temperature should be assumed and the calculations should be repeated from step 3(i) onward until the assumed temperature and the calculated one are very close.

Step 5
Calculate the overall heat transfer coefficient:

$$U = \frac{1}{A \sum R} = \text{..}$$

Step 6
Calculate the heat losses:

$$q = AU\Delta T_{overall} = \text{..}$$

Exercise 8.7

Run the spreadsheet program *Convection 1.xls*. See how the value of h_o varies when the outside wall surface temperature is adjusted in the cell F34. Adjust the spreadsheet program *Convection 1.xls* to calculate the heat losses from the side, the upper plane, and the lower plane of the dryer of Exercise 8.6. Compare the results.
Use the following empirical correlations for the physical properties of air as a function of temperature (in °C):

$$\text{Density } \rho = 1.284 - 3.9 \times 10^{-3}T + 6.05 \times 10^{-6}T^2 \, (kg/m^3)$$

$$\text{Viscosity } \mu = 1.75 \times 10^{-5} + 4.17 \times 10^{-8}T (Pas)$$

$$\text{Heat capacity } c_{p_t} = 1004.1 + 4.28 \times 10^{-2}T + 3 \times 10^{-4}T^2 (J/kg\,°C)$$

$$\text{Thermal conductivity } k = 2.441 \times 10^{-2} + 7.12 \times 10^{-5}T (W/m°C)$$

$$\text{Volumetric expansion } \beta = 3.6 \times 10^{-3} - 1.1 \times 10^{-5}T + 1.66 \times 10^{-8}T^2 (1/K)$$

(*Hints*: Find suitable equations from the literature (e.g., Ref. 2, 3, 4) to calculate the Nu number for a heated plate facing upward (top side of the dryer) and for a heated plate facing downward (bottom side of the dryer). Introduce the constants in the appropriate cells. Write the equations for the overall heat transfer coefficient and the heat transfer rate. Run the program.)

Exercise 8.8

Get familiar with the spreadsheet program *Convection 2.xls*. Run the program. See how the outside surface temperature of the vertical wall of the dryer of Exercise 8.6 and the heat transfer coefficient h_o vary when the inside air temperature of the dryer varies from 50 °C to 200 °C. Adjust the spreadsheet and plot how the heat losses through the vertical wall are affected when the inside air temperature of the dryer varies from 50 °C to 200 °C.

Exercise 8.9

A loaf of bread with dimensions 0.25 m x 0.10 m x 0.10 m is exposed to room temperature as it exits from the oven. Develop a spreadsheet program to calculate the heat transfer coefficient on the top horizontal surface of the bread as it cools down from 150 °C to 35 °C. Plot the heat transfer coefficient vs. the bread surface temperature. Assume a room temperature of 21 °C and uniform bread surface temperature. (*Hints:* Find a suitable equation from the literature (e.g., Ref. 2, 3, 4) to calculate the Nusselt number for a heated plate facing upward. Use as the characteristic dimension the arithmetic mean of the dimensions of the rectangle. Use the empirical correlations given in Exercise 8.7 for the physical properties of the air. Calculate Nu and then h. Use the IF function in the cell that gives the bread surface temperature and ITERATE to reduce the bread surface temperature from 150 °C to 35 °C.)

Exercise 8.10

A liquid food is heated at a rate of 1 kg/s in a double-pipe heat exchanger. The liquid enters the inside tube at 10 °C and exits at 70 °C. Water is used as the heating medium, entering the annular space of the heat exchanger at 90 °C, flowing in a countercurrent mode, and exiting at 60 °C. If the overall heat transfer coefficient of the heat exchanger is 200 W/m2 °C, calculate the required water flow rate, the required heat transfer area, and the effectiveness of the heat exchanger. Use a value of 3.5 kJ/kg °C for the heat capacity of the liquid and 4.18 kJ/kg °C for the water.

Solution

Step 1
Draw the process diagram:

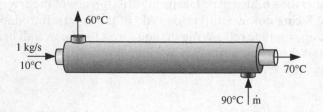

Step 2
State your assumptions:...

Step 3
Write an enthalpy balance on the heat exchanger and calculate the mass flow rate of water:

....

$$\dot{m} = \text{.......................}$$

Step 4
Calculate the logarithmic mean temperature difference:

$$\Delta T_{LM} = \text{..}$$

Step 5
Calculate the heat transfer area from the heat transfer rate equation:

$$A = \frac{q}{U\Delta T_{LM}} = \text{..}$$

Step 6
Calculate the effectiveness of the heat exchanger:

i) The effectiveness of a heat exchanger is given by (see, for example, Ref. 4):

$$\varepsilon = \frac{C_{hot}(T_{hot\ in} - T_{hot\ out})}{C_{min}(T_{hot\ in} - T_{cold\ in})} \qquad (8.5)$$

or

$$\varepsilon = \frac{C_{cold}(T_{cold\ out} - T_{cold\ in})}{C_{min}(T_{hot\ in} - T_{cold\ in})} \qquad (8.6)$$

where

$$C_{hot} = \dot{m}_{hot}c_{p\ hot}$$

$$C_{cold} = \dot{m}_{cold}c_{p\ cold}$$

ii) Calculate C_{hot} and C_{cold} and determine which one is C_{min}:

$$C_{hot} = \text{..}W/°C$$

$$C_{cold} = \text{..}W/°C$$

iii) Substitute these values into eqn (8.5) or eqn (8.6) and calculate ε:

ε = ... = ...

Exercise 8.11

If the liquid of Exercise 8.10 is heated in a 1-2 multiple pass shell and tube heat exchanger, calculate the mean temperature difference and the area of the heat exchanger. The overall heat transfer coefficient is $250W/m^{2}°C$. Use the same values for the temperature of the liquid food and water.

Solution

Step 1
Draw the process diagram:

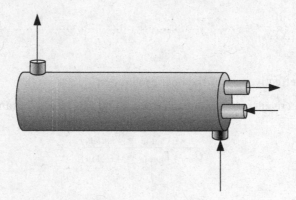

Step 2
Calculate the mean temperature difference:
The logarithmic mean temperature difference found for the 1-1 heat exchanger of Exercise 8.10 must be corrected since the flow is mixed (cocurrent and countercurrent flow).

i) Calculate the dimensionless ratios:

$$\frac{T_{hi} - T_{ho}}{T_{co} - T_{ci}} = \frac{90 - 60}{70 - 10} = 0.5$$

and

$$\frac{T_{co} - T_{ci}}{T_{hi} - T_{ci}} = \text{..}$$

ii) Find the correction factor from an appropriate diagram from the literature (e.g., Ref. 3, 4). With the above values of the dimensionless ratios, the correction factor for a 1-2 heat exchanger is 0.55. Therefore the corrected mean temperature difference is:

$$\Delta T_m = 0.55 \cdot \Delta T_{LM} = \text{...}$$

Step 3
Calculate the heat transfer area:

$$A = \frac{\text{...}}{\text{...}} = \text{...}$$

Exercise 8.12

Calculate the mean temperature difference and the area of the heat exchanger of Exercise 8.11 if saturated steam at 101325 Pa absolute pressure is used as the heating medium instead of water. Use the same values for the temperature of the liquid food. Assume that the overall heat transfer coefficient in this case is $300 \text{W}/\text{m}^{2}{}^\circ\text{C}$.

Solution

Step 1
Calculate the mean temperature difference:
The temperature of the heating medium is: ....

The correction factor for the mean temperature difference for a multiple-pass heat exchanger when one of the fluids undergoes a phase change is 1.
Therefore:

$$\Delta T_m = \text{...}$$

Step 2
Calculate the heat transfer area:

$$A = \text{..}$$

Exercise 8.13

Milk is pasteurized at a rate of 1.5kg/s in a plate heat exchanger that consists of a heat regeneration section, a heating section, and a cooling section. The milk enters the regeneration section at 5°C and exits at 45°C. It then enters the heating section where it is heated up to 72°C, flows through the holder, and returns to the regeneration section where it is cooled to 32 °C. From the regenerator, it flows to the cooling section where it is cooled down to 5 °C. Calculate 1) the required heat transfer area of each section, 2) the required flow rate of the heating water in the heating section, and 3) the brine exit temperature in the cooling section. It is given that the streams in each section flow in countercurrent mode; the heating water enters the heating section at 90 °C and exits at 80 °C; the brine enters the cooling section at –5 °C at a flow rate of 1kg/s; the overall heat transfer coefficients are 1100W/m²°C for the regenerator, 1300W/m²°C for the heating section, and 800W/m²°C for the cooling section; the heat capacities throughout the process are 3.9kJ/kg°C for the milk, 4.19kJ/kg°C for the water, and 3.5kJ/kg°C for the brine.

Solution

Step 1
Draw the process diagram:

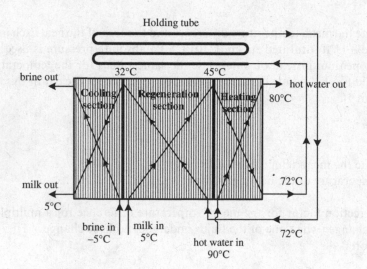

Step 2
Calculate the heat transfer area in the regenerator.
  i) Calculate the heat transferred from the hot stream to the cold stream:

$$q_R = \dot{m}_m c_p (T_{hot\ in} - T_{hot\ out}) = \ldots\ldots\ldots\ldots\ldots\ldots\ldots\ldots\ldots\ldots$$

  ii) Calculate the driving force for heat transfer (mean temperature difference ΔT_m). Since

$$\Delta T_1 = \Delta T_2 = \ldots\ldots\ldots\ldots\ldots\ldots\ldots\ldots\ldots\ldots\ldots\ldots\ldots\ldots\ldots\ldots$$

$$\Delta T_m = \Delta T_1 = \Delta T_2 = \ldots\ldots\ldots\ldots\ldots\ldots\ldots\ldots\ldots\ldots\ldots\ldots$$

iii) Calculate the heat transfer area of the regenerator:

$$A_R = \frac{q_R}{U_R \Delta T_m} = \text{...}$$

Step 3
Calculate the heat transfer area in the heating section.

i) Calculate the mass flow rate of water. Write an enthalpy balance around the heating section and calculate m_{water}.

....

ii) Calculate the heat transfer area of the heat exchanger in the heating section (work as for the regenerator):

....

....

....

Step 4
Calculate the heat transfer area in the cooling section.

i) Write an enthalpy balance around the cooling section and calculate the brine exit temperature:

....

ii) Calculate the heat transfer area of the heat exchanger in the cooling section (work as for the regenerator):

....

....

....

Step 5
Write an enthalpy balance in the heating section and calculate the mass flow
rate of the heating water:

....

$$\dot{m}_w = ...$$

Step 6
Calculate the brine exit temperature:

....

Exercise 8.14

In the previous exercise, calculate: 1) the necessary length of the holding tube so
that at steady state the fastest moving liquid remains in the holding tube for 16 s;
and 2) the required thickness of insulation so that the temperature of the milk
drops by 0.1 °C in the holding tube. The following data are given: the mean
velocity of the liquid is $0.80 \, v_{max}$, the heat transfer coefficient on the milk side is
$1500 \, W/m^2 \, °C$ and on the airside is $5 \, W/m^2 \, °C$, the thermal conductivity of the
pipe is $15 \, W/m \, °C$ and of the insulation is $0.05 \, W/m \, °C$, the room temperature is
$20 \, °C$, the inside diameter of the pipe is 0.0356 m, the thickness of the pipe wall is
1.2 mm, and the density of the milk is $1010 \, kg/m^3$.

Solution

Step 1

    i) Calculate the mean velocity:

$$v = ...m/s$$

    ii) Calculate the maximum velocity:

$$v_{max} = \frac{v}{0.8} = ...m/s$$

    iii) Calculate the required length:

$$L = v_{max}t = ...$$

Step 2
Develop a spreadsheet program to calculate the thickness of the insulation by
iteration.

    i) Calculate the heat that must be removed so that the temperature of the milk drops by 0.1 °C.

    ii) Select a value for the insulation thickness.

    iii) Calculate the overall heat transfer coefficient.

    iv) Calculate heat losses with the heat transfer rate equation.

    v) If heat losses calculated in iv) are higher than those calculated in i), increase the thickness of the insulation and repeat the calculations from step ii). ITERATE to increase the thickness of insulation until the heat losses calculated in iv) are equal to the heat calculated in i). Use the IF function in the cell that gives the insulation thickness.

Exercise 8.15

The overall heat transfer coefficient based on the outside surface area of the tubes of a sugar evaporator decreased during operation from $1200 W/m^2 °C$ to $800 W/m^2 °C$ due to fouling. Calculate the fouling coefficient.

Solution

Step 1
State your assumptions: The effect of fouling on the heat transfer area A_o is negligible.

Step 2
By definition:

$$U_o = \frac{1}{A_o \sum R} \text{ or } \sum R = \frac{1}{U_o A_o}$$

and since resistances are additive:

$$\sum R_{final} = \sum R_{initial} + R_{fouling} \qquad (8.7)$$

where

$$R_{fouling} = \frac{1}{h_{fouling} A_o}$$

Step 3
Substitute into eqn (8.7) and calculate $h_{fouling}$:

$$h_{fouling} = \dots\dots\dots\dots\dots\dots\dots\dots\dots\dots\dots\dots\dots\dots\dots\dots\dots\dots\dots\dots W/m^2 °C$$

Exercise 8.16

A fouling coefficient of $1990 W/m^2 °C$ is recommended for use in designing heat transfer equipment for vegetable oils (see for example Ref. 3, p. 187).

How much more heat transfer area will be required in a heat exchanger to compensate for the reduction in the overall heat transfer coefficient due to this fouling? The overall heat transfer coefficient based on the inside area of the tubes for the clean heat exchanger is 300 W/m^2 °C.

Solution

Step 1
State your assumptions

- The effect of fouling on the diameter of the tubes is negligible.
- The heat transfer driving force will be the same in the clean and the fouled heat exchangers.
- The heat transferred will be the same in the clean and the fouled heat exchangers.

Step 2

i) As shown in Exercise 8.15:

$$\frac{1}{U_{total}} = \frac{1}{U_{clean}} + \frac{1}{h_{foul}}$$

ii) Calculate U_{total}:

$$U_{total} = \text{...}$$

Step 3
The required heat transfer area for the fouled heat exchanger is:

$$A_{foul} = \frac{q}{U_{total}\Delta T_{LM}} \tag{8.8}$$

Step 4
The required heat transfer area for the clean heat exchanger is:

$$A_i = \frac{q}{U_{clean}\Delta T_{LM}} \tag{8.9}$$

Step 5
Divide eqn (8.8) by eqn (8.9), substitute values and calculate the increase in heat transfer area:

$$\frac{A_{foul}}{A_i} = \text{...}$$

Chapter 9
Heat Transfer by Radiation

Review Questions

Which of the following statements are true and which are false?

1. The presence of matter between the points that exchange energy by radiation is not necessary.
2. A black body absorbs all the visible light but may reflect other wavelengths of the incident radiation.
3. The radiation emitted by surfaces at temperatures less than about 800 K is not visible by the human eye.
4. A metal rod turns red when heated in a fire because it reaches a temperature at which it emits radiation visible to the human eye.
5. A leaf is seen as green because it emits radiation at the green wave length.
6. A surface is seen as white because it reflects all the wavelengths of visible light.
7. The Stefan-Boltzman law states that the total radiation from a black body is proportional to the square of the absolute temperature of the body.
8. In heat transfer calculations involving radiation, the temperature is usually expressed in degrees Celsius.
9. Emissivity is defined as the ratio of the radiation emitted by a surface to that emitted by a black body at the same temperature.
10. Absorptivity is defined as the ratio of the radiation absorbed by a surface to that absorbed by a black body.
11. When a surface is at thermal equilibrium with its surroundings, its emissivity is equal to its absorptivity.
12. Rough surfaces have lower emissivity than polished surfaces.
13. The shape factor or view factor is equal to the fraction of radiation leaving area A_1 that is intercepted by area A_2.
14. The view factor of a small surface surrounded by a large surface is equal to 0.5.
15. Surfaces with a high emissivity to absorptivity ratio are suitable for collecting solar energy.

S. Yanniotis, *Solving Problems in Food Engineering.*
© Springer 2008

Examples

Example 9.1

A horizontal steel pipe with 2 in outside diameter has an outside surface temperature of 150 °C. Calculate the heat losses per m length of pipe, taking into account convection and radiation losses. Assume that the emissivity of the surface of the tube is 0.75, the pipe is exposed to a room temperature of 20 °C, and the convective heat transfer coefficient is 15 W/m$^{2\circ}$C.

Solution

Step 1
Calculate radiation losses:

i) Assumptions:

- The pipe is considered a small grey body in a large enclosure.
- The temperature of the surfaces seen by the pipe is equal to the air temperature.

ii) The net heat transferred from the pipe to the surroundings by radiation is:

$$q_{12} = A_1 F_{12} \sigma (T_1^4 - T_2^4) \qquad (9.1)$$

with

$$F_{12} = \frac{1}{f_{12} + \left(\frac{1}{\varepsilon_1} - 1\right) + \frac{A_1}{A_2}\left(\frac{1}{\varepsilon_2} - 1\right)}$$

Because all the radiation from A_1 (the pipe) is intercepted by A_2 (the surrounding walls), $f_{12} = 1$. Also, because the surface area of the pipe (A_1) is negligible compared to the surface area of the surroundings, $A_1/A_2 \approx 0$. Therefore $F_{12} = \varepsilon_1$.

Equation (9.1) then becomes:

$$q_{12} = A_1 \varepsilon_1 \sigma (T_s^4 - T_a{}^4) \qquad (9.2)$$

Step 2
Calculate convection losses:

$$q_c = h_c A_1 (T_s - T_a) \qquad (9.3)$$

Step 3
Calculate total losses:

$$q_{\text{toatl}} = q_c + q_r = (h_c + h_r) A_1 (T_s - T_a) \qquad (9.4)$$

where

$$h_r = \frac{\varepsilon_1 \sigma \left(T_s^4 - T_a^4\right)}{T_s - T_a} = \frac{(0.75)\left(5.676 \times 10^{-8}\,\text{W/m}^2\text{K}^4\right)\left(423^4 - 293^4\right)\text{K}^4}{(423 - 293)\text{K}}$$

$$= 8.1\,\text{W/m}^2\text{K}$$

Remember that temperature in radiation problems must be in degrees Kelvin.

Step 4
Substitute values into eqn (9.4) and calculate q_{total}.

$$q_{total} = (15 + 8.1)\text{W/m}^2\text{K} \times (\pi \times 1 \times 2 \times 0.0254)\text{m}^2 (423 - 293)\text{K} = 479.3\,\text{W}$$

Exercises

Exercise 9.1

A "thermos" is made of two coaxial glass cylinders with the annular space in between evacuated to minimize heat transfer to the environment and keep the contained liquid at a constant temperature (e.g., hot coffee). The glass surfaces are polished, having an emissivity of 0.095. The diameters of the inner and outer cylinders are 10 cm and 11 cm respectively and their length is 20 cm. The temperature of the inner and outer cylinders is 90 °C and 30 °C respectively. Calculate the rate of heat loss by radiation through the cylindrical surface from the inner to the outer cylinder.

Solution

Step 1
Write the equation for heat transfer by radiation:

The net heat transferred from the inner cylinder to the outer by radiation is:

$$q_{12} = A_1 F_{12} \sigma \left(T_1^4 - T_2^4\right)$$

with

$$F_{12} = \frac{1}{f_{12} + \left(\frac{1}{\varepsilon_1} - 1\right) + \frac{A_1}{A_2}\left(\frac{1}{\varepsilon_2} - 1\right)} = \frac{1}{\left(\frac{1}{\varepsilon_1} - 1\right) + \frac{D_1}{D_2}\left(\frac{1}{\varepsilon_2} - 1\right)}$$

(Explain why $f_{12} = 1$ in the above equation).

Step 2
Substitute values and calculate q_{12}:

$q_{12} =$ ....

Exercise 9.2

A cookie with $50\,\mathrm{cm}^2$ surface area, emissivity equal to 0.85, and surface temperature equal to $80\,°\mathrm{C}$ is being baked in a domestic oven. Calculate the heat transfer rate to the cookie by radiation if the walls of the oven are at $180\,°\mathrm{C}$.

Solution

Step 1
State your assumptions: ....

Step 2
Write the equation for heat transfer by radiation:

$$q_{21} = A_2 F_{21}\sigma\left(T_w^4 - T_s^4\right) = A_1 F_{12}\sigma\left(T_w^4 - T_s^4\right)$$

Step 3
Substitute values and calculate q_{21}:

$q_{21} =$ ....

Exercise 9.3

An orange of spherical shape with diameter of 8 cm is hanging on an orange tree during a clear-sky night. The orange gains heat from the air by natural convection and loses heat by radiation. One third of the orange surface "sees" the sky, while the other 2/3 "sees" the leaves of the tree which are at the same temperature as the orange. If the air temperature drops, calculate the air temperature at which the orange will reach the temperature of $-1\,^{\circ}C$, which is the initial freezing point of the orange. Assume a convective heat transfer coefficient of $3\,W/m^2\,^{\circ}C$, emissivity of the orange at 0.8 and sky temperature at 230 K. If a big blower is used to circulate the air around the tree so that the heat transfer coefficient increases to $6\,W/m^2\,^{\circ}C$, what will be the air temperature at which freezing of the orange will start?

Solution

Step 1
State your assumptions:

....

Step 2
Calculate the rate of heat gain by convection:

$q_c = $ ....

Step 3
Calculate the rate of heat loss by radiation:

$$q_r = \frac{A_1}{3} \varepsilon_1 \sigma \left(T_{orange}^4 - T_{sky}^4 \right) = $$ ....

Step 4
Calculate the air temperature by equating the last two equations:

....

Step 5
Repeat the calculations for $h = 6W/m^2\,^{\circ}C$.

Exercise 9.4

A black air duct is used to collect solar energy to heat the air for a solar food dryer. The surface area of the collector exposed to the sun is $10\,m^2$, the temperature at the surface of the collector is 70 °C, the sky temperature is 20 °C, the air temperature is 30 °C, the heat transfer coefficient at the surface of the collector is 2 W/m^2 °C, the absorptivity of the collector to the sun rays is 0.97 and the emissivity is 0.2. Calculate the energy collected if the solar insolation is 700 W/m^2.

Solution

Step 1
State your assumptions:

- The bottom and the sides of the collector are insulated so that heat losses are negligible.
- The system is at steady state.
- ....

Step 2
Write an energy balance on the collector:

Input : directsolarinsolation + skyradiation.

or

$$q_{in} = \alpha A q_{direct\ solar} + \alpha A \sigma T_{sky}^4$$

Output : heat lost by convection to the surroundings + heat lost by

radiation + heat transferred to the air of the dryer.

or

$$q_{out} = hA(T_s - T_{air}) + \varepsilon A \sigma T_s^4 + q_{useful}$$

At steady state $q_{in} = q_{out}$.

Step 3
Substitute values and solve for q_{useful}.

$$q_{useful} = \dots\dots\dots\dots\dots\dots\dots\dots\dots\dots\dots\dots\dots\dots\dots\dots\dots\dots\dots\dots$$

Exercise 9.5

Develop a spreadsheet program to calculate and plot the heat transferred by convection and radiation from the surface of the bread of Exercise 8.9 as the bread cools from 150 °C to 35 °C. Assume that a) the emissivity of bread is 0.9 and b) the bread surface temperature is uniform.

Chapter 10
Unsteady State Heat Transfer

Theory

The governing equation for unsteady state heat transfer in rectangular coordinates in the x-direction is:

$$\frac{\partial T}{\partial t} = \alpha \frac{\partial^2 T}{\partial x^2}$$

where

T = temperature, °C
t = time, s
x = distance from the center plane, m
α = thermal diffusivity, m^2/s

Analytical solutions for various initial and boundary conditions are available in the literature (e.g., Ref. 5 and 6). The solution is of the form:

$$\frac{T_e - T}{T_e - T_o} = \sum_n A_n \exp(B_n Fo)$$

where

T_e = equilibrium temperature (temperature of the environment)
T_o = initial temperature
T = temperature at time t and point x
A_n = constant
B_n = constant
Fo = Fourier number ($= \alpha t / L^2$)

Table 10.1 Solutions of the unsteady state heat transfer equation with uniform initial temperature and constant surface temperature (negligible surface resistance, Bi > 40)

1) Local Temperature

 a) Rectangular coordinates

$$\frac{T_e - T}{T_e - T_o} = \frac{4}{\pi} \sum_{n=0}^{\infty} \frac{(-1)^n}{2n+1} \cos\left(\frac{(2n+1)\pi x}{2L}\right) \exp\left(-\frac{(2n+1)^2 \pi^2}{4} Fo\right)$$

 b) Cylindrical coordinates

$$\frac{T_e - T}{T_e - T_o} = \frac{2}{R} \sum_{n=1}^{\infty} \frac{J_0(r\delta_n)^n}{\delta_n J_1(R\delta_n)} \exp\left(-R^2\delta_n^2 Fo\right)$$

 with δ_n being roots of the Bessel function $J_o(R\delta_n) = 0$.

 The first five roots of $J_o(x)$ are: 2.4048, 5.5201, 8.6537, 11.7915 and 14.9309 (see Table A.1). Consequently,
 $\delta_1 = 2.4048/R$, $\delta_2 = 5.5201/R$, $\delta_3 = 8.6537/R$, $\delta_4 = 11.7915/R$ and $\delta_5 = 14.9309/R$

 c) Spherical coordinates

$$\frac{T_e - T}{T_e - T_o} = -\frac{2R}{\pi r} \sum_{n=1}^{\infty} \frac{(-1)^n}{n} \sin\left(\frac{n\pi r}{R}\right) \exp\left(-n^2\pi^2 Fo\right)$$

2) Mean Temperature

 a) Rectangular coordinates

$$\frac{T_e - T_m}{T_e - T_o} = \frac{8}{\pi^2} \sum_{n=0}^{\infty} \frac{1}{(2n+1)^2} \exp\left(-\frac{(2n+1)^2\pi^2}{4} Fo\right) =$$

$$= \frac{8}{\pi^2}\left(\exp(-2.47 Fo) + \frac{1}{9}\exp(-22.2 Fo) + \frac{1}{25}\exp(-61.7 Fo) + \ldots\right)$$

 b) Cylindrical coordinates

$$\frac{T_e - T_m}{T_e - T_o} = \frac{4}{\pi^2} \sum_{n=1}^{\infty} \frac{\pi^2}{R^2\delta_n^2} \exp\left(-R^2\delta_n^2 Fo\right) =$$

$$= \frac{4}{\pi^2}\left(1.7066\exp(-5.783 Fo) + 0.324\exp(-30.5 Fo)\right.$$

$$\left. + 0.132\exp(-74.9 Fo) + \ldots\right)$$

 c) Spherical coordinates

$$\frac{T_e - T_m}{T_e - T_o} = \frac{6}{\pi^2} \sum_{n=1}^{\infty} \frac{1}{n^2} \exp\left(-n^2\pi^2 Fo\right) =$$

$$= \frac{6}{\pi^2}\left(\exp(-\pi^2 Fo) + \frac{1}{4}\exp(-4\pi^2 Fo) + \frac{1}{9}\exp(-9\pi^2 Fo) + \ldots\right)$$

where

 A = heat transfer surface area, m^2
 α = thermal diffusivity, m^2/s
 Bi = Biot number $(= hR/k)$
 c_p = heat capacity, J/kg °C
 erf = the error function

Table 10.2 Solutions of the unsteady state heat transfer equation for uniform initial temperature and both resistances, surface and internal, significant $(0.1 < Bi < 40)$

1) Local Temperature

   a) Rectangular coordinates

$$\frac{T_e - T}{T_e - T_o} = \sum_{n=1}^{\infty} \frac{2Bi}{\delta_n^2 + Bi^2 + Bi} \frac{\cos\left(\frac{x}{L}\delta_n\right)}{\cos(\delta_n)} \exp\left(-\delta_n^2 Fo\right)$$

    with δ_n roots of: $\delta \tan\delta = Bi$

   b) Cylindrical coordinates

$$\frac{T_e - T}{T_e - T_o} = \sum_{n=1}^{\infty} \frac{2\,Bi}{\delta_n^2 + Bi^2} \frac{J_o\left(\frac{r}{R}\delta_n\right)}{J_o(\delta_n)} \exp\left(-\delta_n^2 Fo\right)$$

    with δ_n roots of: $\delta\,J_1(\delta) = Bi\,J_o(\delta)$

   c) Spherical coordinates

$$\frac{T_e - T}{T_e - T_o} = \frac{R}{r} \sum_{n=1}^{\infty} \frac{2\,Bi}{\delta_n^2 + Bi^2 - Bi} \frac{\sin\left(\frac{r}{R}\delta_n\right)}{\sin(\delta_n)} \exp\left(-\delta_n^2 Fo\right)$$

    with δ_n roots of: $\delta \cot\delta = 1 - Bi$

   For the center of the sphere $(r = 0)$

$$\frac{T_e - T}{T_e - T_o} = \sum_{n=1}^{\infty} \frac{2\,Bi}{\delta_n^2 + Bi^2 - Bi} \frac{\delta_n}{\sin(\delta_n)} \exp\left(-\delta_n^2 Fo\right)$$

2) Mean Temperature

   a) Rectangular coordinates

$$\frac{T_e - T_m}{T_e - T_o} = \sum_{n=1}^{\infty} \frac{2Bi^2}{(\delta_n^2 + Bi^2 + Bi)\delta_n^2} \exp\left(-\delta_n^2 Fo\right)$$

    with δ_n roots of: $\delta \tan\delta = Bi$

   b) Cylindrical coordinates

$$\frac{T_e - T_m}{T_e - T_o} = \sum_{n=1}^{\infty} \frac{4\,Bi^2}{(\delta_n^2 + Bi^2)\,\delta_n^2} \exp\left(-\delta_n^2 Fo\right)$$

    with δ_n roots of: $\delta\,J_1(\delta) = Bi\,J_o(\delta)$

   c) Spherical coordinates

$$\frac{T_e - T_m}{T_e - T_o} = \sum_{n=1}^{\infty} \frac{6\,Bi^2}{(\delta_n^2 + Bi^2 - Bi)\,\delta_n^2} \exp\left(-\delta_n^2 Fo\right)$$

    with δ_n roots of: $\delta \cot\delta = 1 - Bi$

Table 10.3 Unsteady state heat transfer with negligible internal resistance $(0.1 < Bi)$

$$\frac{T_e - T}{T_e - T_o} = \exp\left(-\left(\frac{hA}{c_p m}\right)t\right) = \exp(-Bi\,Fo)$$

$$t = \frac{mc_p}{hA} \ln\frac{T_e - T_o}{T_e - T}$$

---

erfc = the complementary error function defined as erfc = 1 - erf
Fo = Fourier number $(= \alpha t / L^2)$
h = heat transfer coefficient, $W/m^2\,°C$
J_o = Bessel function of the first kind of order zero
J_1 = Bessel function of the first kind of order one
k = thermal conductivity, $W/m^2°C$
L = half thickness of the plate, m
m = mass, kg

Table 10.4 Solutions of the unsteady state heat transfer equation for a semi-infinite body with uniform initial temperature

1) Constant surface temperature

$$\frac{T_e - T}{T_e - T_o} = erf\left(\frac{x}{2\sqrt{\alpha t}}\right)$$

2) Convection at the surface

$$\frac{T_e - T}{T_e - T_o} = erf\left(\frac{x}{2\sqrt{\alpha t}}\right) + \left[exp\left(\frac{hx}{k} + \frac{h^2 \alpha t}{k^2}\right)\right]\left[erfc\left(\frac{x}{2\sqrt{\alpha t}} + \frac{h\sqrt{\alpha t}}{k}\right)\right]$$

where

R = radius of cylinder or sphere, m

r = distance of the point from the centerline of the cylinder or center of the sphere, m

x = distance of the point from the center plane of the plate, or from the surface in the case of semi-infinite body, m

t = time, s

T_e = equilibrium temperature (temperature of the environment or surface temperature)

T_o = initial temperature

T = temperature at time t and point x

T_m = mean temperature

To calculate the temperature as a function of time and position in a solid:

1. Identify the geometry of the system. Determine if the solid can be considered as plate, infinite cylinder, or sphere.
2. Determine if the surface temperature is constant. If not, calculate the Biot number and decide the relative importance of internal and external resistance to heat transfer.
3. Select the appropriate equation.
4. Calculate the Fourier number.
5. Find the temperature by applying the selected equation (if $F_0 > 0.2$, calculate the temperature either using only the first term of the series solution or using Heisler or Gurnie-Lurie charts).

REVIEW QUESTIONS

Which of the following statements are true and which are false?

1. If the temperature at any given point of a body changes with time, unsteady state heat transfer occurs.
2. Thermal diffusivity is a measure of the ability of a material to transfer thermal energy by conduction compared to the ability of the material to store thermal energy.

3. Materials with high thermal diffusivity will need more time to reach equilibrium with their surroundings.
4. The Biot number (Bi) expresses the relative importance of the thermal resistance of a body to that of the convection resistance at its surface.
5. If Bi, the external resistance is negligible.
6. If Bi > 40, the surface temperature may be assumed to be equal to the temperature of the surroundings.
7. If Bi, the internal resistance is significant.
8. If Bi Bi, the temperature of the body may be assumed to be uniform.
9. Problems with Bi are treated with the lumped capacitance method.
10. The Fourier number (Fo) has dimensions of time.
11. The Heisler charts give the temperature at the center of an infinite slab, infinite cylinder, and sphere when $F_O > 0.2$.
12. The Gurney-Lurie charts give the temperature at any point of an infinite slab, infinite cylinder, and sphere when $F_O > 0.2$.
13. Thermal penetration depth is defined as the distance from the surface at which the temperature has changed by 10% of the initial temperature difference.
14. Until the thermal penetration depth becomes equal to the thickness of a finite body heated from one side, the body can be treated as a semi-infinite body.
15. A cylinder of finite length can be treated as an infinitely long cylinder if the two ends of the cylinder are insulated or if its length is at least 10 times its diameter.

Examples

Example 10.1

A steak 2 cm thick is put on a hot metallic plate. The surface of the steak in contact with the hot surface immediately attains a temperature of 120°C and retains this temperature. Calculate the temperature 1.1 cm from the hot surface of the steak after 15 min, if the initial temperature of the meat is 5°C and the thermal diffusivity of the meat is $1.4 \times 10^{-7} \mathrm{m^2/s}$.

Solution

Step 1
Draw the process diagram:

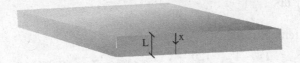

Step 2
Identify the geometry of the system.
The shape of the steak is a plate heated from one side.

Step 3
Examine the surface temperature.
The surface temperature is constant.

Step 4
Select the appropriate equation.
The equation for a plate with uniform initial temperature and constant surface temperature is (Table 10.1):

$$\frac{T_e - T}{T_e - T_o} = \frac{4}{\pi} \sum_{n=0}^{\infty} \frac{(-1)^n}{2n + 1} \cos\left(\frac{(2n + 1)\pi x}{2L}\right) \exp\left(-\frac{(2n + 1)^2 \pi^2}{4} Fo\right) \quad (10.1)$$

Step 5
Calculate the Fourier number.
Since the steak is heated *only from one side*, the characteristic dimension is the thickness of the steak, *not the half thickness*. Therefore:

$$Fo = \frac{\alpha t}{L^2} = \frac{(1.4 \times 10^{-7}\,\text{m}^2/\text{s})(15 \times 60)}{0.02^2\,\text{m}^2} = 0.315$$

Step 6
Calculate the temperature
Since $F_O > 0.2$, only the first term of the sum in eqn (10.1) can be used without appreciable error. Alternatively, the solution can be read directly from a Gurney-Lurie chart (see for example Ref. 3), which gives the dimensionless ratio $(T_e - T)/(T_e - T_o)$ vs. the Fourier number. The temperature below is calculated both ways:

   i) From the equation:

$$\frac{T_e - T}{T_e - T_o} = \frac{4}{\pi} \sum_{n=0}^{\infty} \frac{(-1)^n}{2n + 1} \cos\left(\frac{(2n + 1)\pi x}{2L}\right) \exp\left(-\frac{(2n + 1)^2 \pi^2}{4} Fo\right)$$

$$\approx \frac{4}{\pi} \frac{(-1)^0}{2 \times 0 + 1} \cos\left(\frac{(2 \times 0 + 1)\pi x}{2L}\right) \exp\left(-\frac{(2 \times 0 + 1)^2 \pi^2}{4} Fo\right)$$

$$= \frac{4}{\pi} \cos\left(\frac{\pi x}{2L}\right) \exp\left(-\frac{\pi^2}{4} Fo\right) = \frac{4}{\pi} \cos\left(\frac{\pi \times 0.009}{2 \times 0.02}\right) \exp\left(-\frac{\pi^2}{4} 0.315\right)$$

$$= 0.445$$

Calculate the temperature as:

$$\frac{T_e - T}{T_e - T_o} = \frac{120 - T}{120 - 5} = 0.445 \rightarrow T = 68.8°C$$

ii) From the Gurnie-Lurie chart:

a) Find the value of Fo = 0.315 on the x-axis of the Gurney-Lurie chart.
b) Find the curve with k/hL = 0 (constant surface temperature or Bi → ∞) and x/L = 0.009/0.02 = 0.45.
c) Read the dimensionless temperature on the y-axis as $(T_e - T)/(T_e - T_o) = 0.45$.
d) Calculate the temperature as

$$\frac{T_e - T}{T_e - T_o} = \frac{120 - T}{120 - 5} = 0.45 \rightarrow T = 68.3°C$$

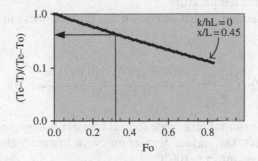

Comment: The reading from the chart can only be approximate

Example 10.2

A hot dog 1.5 cm in diameter and 16 cm in length with 5°C initial temperature is immersed in boiling water. Calculate a) the temperature 3 mm under the surface after 2 min, b) the temperature at the center after 2 min, c) the time necessary to reach 81°C at the center, d) the average temperature of the hot dog after 2 min, and e) the heat that will be transferred to the hot dog during the 2 min of boiling. Assume that the thermal conductivity is 0.5W/m °C, the density is 1050kg/m³, the heat capacity is 3.35kJ/kg °C, and the heat transfer coefficient at the surface of the hot dog is 3000W/m² °C.

Solution

Step 1
Draw the process diagram:

Step 2
Identify the geometry of the system.
The hot dog has a cylindrical shape with L/D equal to:

$$\frac{L}{D} = \frac{0.16\,m}{0.015\,m} = 10.7$$

Since L/D > 10, the hot dog can be treated as an infinite cylinder. The contribution of heat transferred through the bases of the cylinder can be neglected.

Step 3
Examine the surface temperature.
Since the surface temperature is unknown, the Biot number must be calculated:

$$Bi = \frac{hR}{k} = \frac{\left(3000\,W/m^2\,^\circ C\right)(0.0075\,m)}{0.5\,W/m^\circ C} = 45$$

Since Bi > 40 the external resistance to heat transfer is negligible. Therefore, it can be assumed that the surface temperature will immediately reach the temperature of the environment, 100°C.

Step 4
Select the appropriate equation to use. The solution for constant surface temperature will be applied (Table 10.1):

$$\frac{T_e - T}{T_e - T_o} = \frac{2}{R}\sum_{n=1}^{\infty}\frac{J_0(r\delta_n)^n}{J_1(R\delta_n)}\exp\left(-R^2\delta_n^2 Fo\right) \qquad (10.2)$$

Step 5
Calculate the Fourier number.
Calculate the thermal diffusivity first:

$$\alpha = \frac{k}{\rho c_p} = \frac{0.5\,W/m^\circ C}{\left(1050\,kg/m^3\right)(3350\,J/kg^\circ C)} = 1.42 \times 10^{-7} m^2/s$$

Then:

$$Fo = \frac{\alpha t}{R^2} = \frac{(1.42 \times 10^{-7}\,m^2/s)(120\,s)}{0.0075^2\,m^2} = 0.30$$

Since Fo > 0.2, only the first term of the series solution can be used without appreciable error. Alternatively, the solution can be read directly from a chart such as Gurney-Lurie or Heisler (see Fig A.3 and A.6 in the Appendix), which gives the dimensionless ratio $(T_e - T)/(T_e - T_o)$ vs. the Fourier number.

1) Calculate the temperature 3 mm under the surface.

    i) From eqn (10.2) above:

Step 1
Find the values of δn.

        Since $Fo > 0.2$ the first term of the series solution is enough. However, two terms will be used in this problem for demonstration purposes. Since $x = R\delta n$ and $R = 0.0075$, the values of δn will be (see Table A.1 in the Appendix):

x	δn
2.4048	320.6
5.5201	736.0

Step 2
Substitute these values into eqn (10.2) and calculate T:

$$\frac{T_e - T}{T_e - T_o} = \frac{2}{R} \sum_{n=1}^{\infty} \frac{J_0(r\delta_n)^n}{\delta_n J_1(R\delta_n)} \exp(-R^2\delta_n^2 Fo)$$

$$\approx \frac{2}{0.0075} \left(\frac{J_0(0.0045 \times 320.6)}{320.6 \times J_1(2.4048)} \exp(-(2.4048^2)(0.30)) \right.$$

$$\left. + \frac{J_0(0.0045 \times 736)}{736 \times J_1(5.5201)} \exp(-(5.5201^2)(0.30)) \right)$$

$$= 266.67 \left(\frac{J_0(1.4427)}{320.6 \times J_1(2.4048)} \exp(-1.7349) \right.$$

$$\left. + \frac{J_0(3.312)}{736 \times J_1(5.5201)} \exp(-9.141) \right)$$

$$= 266.67 \left(\frac{0.5436}{320.6 \times 0.5192} \exp(-1.7349) \right.$$

$$\left. + \frac{-0.3469}{736 \times (-0.3403)} \exp(-9.141) \right)$$

$$= 266.67(5.76 \times 10^{-4} + 1.48 \times 10^{-7}) = 0.154$$

or

$$\frac{T_e - T}{T_e - T_o} = \frac{100 - T}{100 - 5} = 0.154 \rightarrow T = 85.4°C$$

Comment: Notice that the 2nd term (1.48×10^{-7}) is much smaller than the first term (5.76×10^{-4}) and could have been neglected.

ii) From the Gurney-Lurie chart for an infinite cylinder (see Fig A.3 in the Appendix):

Step 1

Calculate the ratio r/R which gives the position of the point (r is the distance of the point of interest from the axis of the cylinder):

$$\frac{r}{R} = \frac{0.0045\,m}{0.0075\,m} = 0.6$$

Step 2

Select the group of curves on the Gurney-Lurie chart that will be used based on the k/hR value.

Since Bi > 40, the curves with k/hR = 0 on the Gurney-Lurie chart can be used.

Step 3

On the Gurney-Lurie chart:

- Find the value of Fo = 0.3 on the x-axis
- Find the curve with k/hR = 0 and r/R = 0.6
- Read the dimensionless temperature on the y-axis as $(T_e - T)/(T_e - T_o) = 0.15$
- Calculate the temperature as

$$\frac{T_e - T}{T_e - T_o} = \frac{100 - T}{100 - 5} = 0.15 \rightarrow T = 85.8°C$$

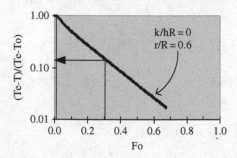

2) Calculate the temperature at the center.

i) From the equation:
As above but with r = 0 :

$$\frac{T_e - T}{T_e - T_o} = \frac{2}{R} \sum_{n=1}^{\infty} \frac{J_0(r\delta_n)^n}{\delta_n J_1(R\delta_n)} \exp(-R^2\delta_n^2 Fo)$$

$$\approx \frac{2}{0.0075} \left(\frac{J_o(0 \times 320.6)}{320.6 \times J_1(2.4048)} \exp(-(2.4048^2)(0.30)) \right.$$

$$+ \frac{J_o(0 \times 736)}{736 \times J_1(5.5201)} \exp(-(5.5201^2)(0.30)) \right)$$

$$= 266.67 \left(\frac{J_o(0)}{320.6 \times J_1(2.4048)} \exp(-1.7349) + \frac{J_o(0)}{736 \times J_1(5.5201)} \exp(-9.141) \right)$$

$$= 266.67 \left(\frac{1}{320.6 \times 0.5192} \exp(-1.7349) + \frac{1}{736 \times (-0.3403)} \exp(-9.141) \right)$$

$$= 266.67(1.06 \times 10^{-3} - 4.279 \times 10^{-7}) = 0.283$$

or

$$\frac{T_e - T}{T_e - T_o} = \frac{100 - T}{100 - 5} = 0.283 \rightarrow T = 73.1°C$$

Comment: Notice that the 2nd term (4.279×10^{-7}) is much smaller than the first term (1.06×10^{-3}) and could have been neglected.

ii) From the Heisler chart for an infinite cylinder (Fig A.6):
   On the Heisler chart:

   - Find the value of Fo = 0.3 in the x-axis.
   - Find the curve with k/hR = 0.
   - Read the dimensionless temperature on the y-axis as $(T_e - T)/(T_e - T_o) = 0.27$
   - Calculate the temperature as

$$\frac{T_e - T}{T_e - T_o} = \frac{100 - T}{100 - 5} = 0.27 \rightarrow T = 74.4°C$$

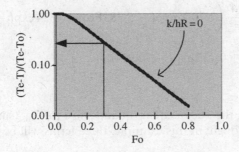

3) Calculate the time necessary to reach 81°C at the center:
   Calculate the dimensionless temperature

$$\frac{T_e - T}{T_e - T_o} = \frac{100 - 81}{100 - 5} = 0.2$$

On the Heisler chart:

- Find on the y-axis the value $(T_e - T)/(T_e - T_o) = 0.2$
- Find the curve $k/hR = 0$.
- Read the Fourier number on the x-axis as Fo = 0.36.
- Find the time from the Fourier number as:

$$t = Fo\,\frac{R^2}{\alpha} = 0.36\,\frac{0.0075^2\,m^2}{1.42 \times 10^{-7}\,m^2/s} = 143 \;\; s$$

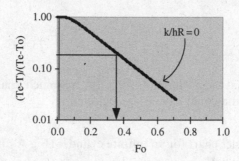

4) Calculate the average temperature.

Step 1
Write the equation for the mean temperature of an infinite cylinder with
   Bi > 40 (Table 10.1):

$$\frac{T_e - T_m}{T_e - T_o} = \frac{4}{\pi^2} \sum_{n=1}^{\infty} \frac{\pi^2}{R^2\delta_n^2}\, \exp\left(-R^2\delta_n^2 Fo\right)$$

$$= \frac{4}{\pi^2}\left(1.7066\exp(-5.783 Fo) + 0.324\exp(-30.5 Fo)\right.$$

$$\left.+ 0.132\exp(-74.9 Fo) + ...\right)$$

Step 2
Substitute values in the above equation and solve for T_m (since Fo > 0.2, the
first term only is enough; in this example, three terms will be used for demon-
stration purposes):

$$\frac{T_e - T_m}{T_e - T_o} = \frac{4}{\pi^2}\left(1.7066 \exp(-5.783 \times 0.3) + 0.324 \exp(-30.5 \times 0.3)\right.$$

$$\left. + 0.132 \exp(-74.9 \times 0.3) + ...\right)$$

$$= \frac{4}{\pi^2}\left(0.3011 + 3.44 \times 10^{-5} + 2.3 \times 10^{-11} + ...\right) = 0.122$$

and

$$T_m = T_e - 0.122(T_e - T_o) = 100 - 0.122(100 - 5) = 88.4\,°C$$

Comment: Notice that the 2nd and 3rd terms of the sum are negligible compared to the 1st term and they could have been omitted.

5) Calculate the heat transferred to the solid in 2 min.
   Because

$$\frac{q_t}{q_e} = \frac{mc_p(T_m - T_o)}{mc_p(T_e - T_o)} = \frac{T_m - T_o}{T_e - T_o} = 1 - \frac{T_e - T_m}{T_e - T_o}$$

the heat transferred to the hot dog in 2 min will be:

$$q_t = mc_p(T_e - T_o)\left(1 - \frac{T_e - T_m}{T_e - T_o}\right)$$

$$= (V\rho)c_p(T_e - T_o)\left(1 - \frac{T_e - T_m}{T_e - T_o}\right)$$

$$= \left(\frac{\pi(0.015^2)\,m^2}{4}(0.16\,m)\left(1050\,\frac{kg}{m^3}\right)\right)\left(3350\,\frac{J}{kg\,°C}\right)(100 - 5\,°C)(1 - 0.122)$$

$$= 8296\,J$$

Example 10.3

12mm × 16mm × 14mm rectangular fruit pieces are immersed in syrup. Calculate the temperature at the center of a piece of fruit after 5 min if the initial temperature of the fruit piece is 20°C, the syrup temperature is 100°C, the heat transfer coefficient at the surface of the fruit piece is 83 W/m^2°C, and the physical properties of the fruit piece are k = 0.5 W/m°C, c_p = 3.8 kJ/kg°C, and ρ = 900 kg/m^3. Assume the fruit piece does not exchange matter with the syrup.

Solution

Step 1
Draw the process diagram:

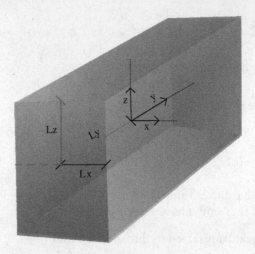

The temperature at the center is affected by the heat transferred from the three directions x, y, and z. The contribution from each direction has to be calculated separately, and the combined effect will be calculated at the end.

1) Heat transferred in the x-direction:

Step 1
Select the appropriate equation to use.

   i) Calculate the Biot number for the x-direction:

$$Bi_x = \frac{hL_x}{k} = \frac{\left(83\,W/m^2\,°C\right)(0.006\,m)}{0.5\,W/m°C} = 1$$

   ii) Since $0.1 < Bi < 40$ both external and internal resistances are important. Therefore, the solution will be (Table 10.2):

$$\frac{T_e - T_x}{T_e - T_o} = \sum_{n=1}^{\infty} \frac{2Bi_x}{\delta_n^2 + Bi_x + Bi_x} \frac{\cos\left(\frac{X}{L_x}\delta_n\right)}{\cos(\delta_n)} \exp\left(-\delta_n^2 Fo_x\right)$$

Step 2
Calculate the Fourier number Fo_x for the x-direction:

   i) Calculate the thermal diffusivity:

$$\alpha = \frac{k}{\rho c_p} = \frac{0.5\,W/m\,°C}{\left(900\,kg/m^3\right)(3800\,J/kg\,°C)} = 1.46 \times 10^{-7} m^2/s$$

ii) Calculate the Fourier number:

$$Fo_x = \frac{\alpha t}{L_x^2} = \frac{(1.46 \times 10^{-7}\,\text{m}^2/\text{s})(300\,\text{s})}{0.006^2\,\text{m}^2} = 1.217$$

Since $Fo_x > 0.2$, only the first term of the sum in the above equation can be used without appreciable error. Alternatively, the solution can be read directly from the Heisler chart.

Step 3
Calculate the temperature:

i) From the equation:
The first root of the equation $\delta \tan\delta = Bi$ for $Bi = 1$ is: $\delta_1 = 0.8603$ (see Table A.2 in the Appendix).

$$\frac{T_e - T_x}{T_e - T_o} \approx \frac{2Bi_x}{\delta_n^2 + Bi_x^2 + Bi_x}\frac{\cos\left(\frac{x}{L_x}\delta_n\right)}{\cos(\delta_n)}\exp\left(-\delta_n^2 Fo_x\right) =$$

$$= \frac{2 \times 1}{0.8603^2 + 1^2 + 1}\frac{\cos\left(\frac{0}{0.006} \times 0.8603\right)}{\cos(0.8603)}\exp\left(-0.8603^2 \times 1.217\right)$$

$$= 0.455$$

ii) From the Heisler chart for an infinite slab:
- Find the value of Fo = 1.22 on the x-axis.
- Find the curve with $k/hL = 1$.
- Read the dimensionless temperature on the y-axis as:

$$\frac{T_e - T_x}{T_e - T_o} = 0.46$$

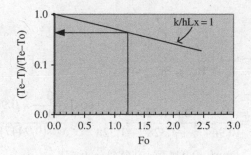

2) Heat transferred in the y-direction:

Step 1
Select the appropriate equation to use.

  i) Calculate the Biot number for the y-direction:

$$Bi_y = \frac{hL_y}{k} = \frac{\left(83\,\text{W/m}^2\,^\circ\text{C}\right)(0.008\,\text{m})}{0.5\,\text{W/m}^\circ\text{C}} = 1.328$$

  ii) Since $0.1 < Bi < 40$ both external and internal resistances are important. Therefore, the solution will be:

$$\frac{T_e - T_y}{T_e - T_o} = \sum_{n=1}^{\infty} \frac{2Bi_y}{\delta_n^2 + Bi_y + Bi_y}\frac{\cos\left(\dfrac{y}{L_y}\delta_n\right)}{\cos(\delta_n)}\exp\left(-\delta_n^2 Fo_y\right)$$

Step 2
Calculate the Fourier number Fo_y for the y-direction:

$$Fo_y = \frac{\alpha t}{L_y^2} = \frac{(1.46 \times 10^{-7}\,\text{m}^2/\text{s})(300\,\text{s})}{0.008^2\,\text{m}^2} = 0.684$$

Since $Fo_y > 0.2$, only the first term of the sum in the above equation can be used without appreciable error. Alternatively, the solution can be read directly from the Heisler chart.

Step 3
Calculate the temperature:

  i) From the equation:
     The first root of the equation $\delta\tan\delta = Bi$ for $Bi = 1.328$ is: $\delta_1 = 0.9447$ (see Table A.2).

$$\frac{T_e - T_y}{T_e - T_o} \approx \frac{2Bi_y}{\delta_n^2 + Bi_y^2 + Bi_y}\frac{\cos\left(\dfrac{y}{L_y}\delta_n\right)}{\cos(\delta_n)}\exp\left(-\delta_n^2 Fo_y\right)$$

$$= \frac{2 \times 1.328}{0.9447^2 + 1.328^2 + 1.328}\frac{\cos\left(\dfrac{0}{0.008} \times 0.9447\right)}{\cos(0.9447)}$$
$$\exp\left(-0.9447^2 \times 0.684\right) = 0.618$$

ii) From the Heisler chart for an infinite slab (Fig. A.5):

- Find the value of Fo = 0.68 on the x-axis.
- Find the curve with k/hL = 0.75 (interpolate between curves 0.7 and 0.8).
- Read the dimensionless temperature on the y-axis as:

$$\frac{T_e - T_y}{T_e - T_o} = 0.61$$

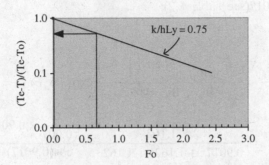

3) Heat transferred in the z-direction:

Step 1
Select the appropriate equation to use.

i) Calculate the Biot number for the z-direction:

$$Bi_z = \frac{hL_z}{k} = \frac{\left(83\,\text{W/m}^2\,{}^\circ\text{C}\right)(0.007\,\text{m})}{0.5\,\text{W/m}^\circ\text{C}} = 1.162$$

ii) Since $0.1 < Bi < 40$ both external and internal resistances are important. Therefore, the solution will be (Table 10.2):

$$\frac{T_e - T_z}{T_e - T_o} = \sum_{n=1}^{\infty} \frac{2Bi_z}{\delta_n^2 + Bi_z + Bi_z} \frac{\cos\left(\frac{z}{L_z}\delta_n\right)}{\cos(\delta_n)} \exp\left(-\delta_n^2 Fo_z\right)$$

Step 2
Calculate the Fourier number Fo_z for the z-direction:

$$Fo_z = \frac{\alpha t}{L_z^2} = \frac{(1.46 \times 10^{-7}\,\text{m}^2/\text{s})(300\,\text{s})}{0.007^2\,\text{m}^2} = 0.894$$

Since $Fo_z > 0.2$, only the first term of the sum in the above equation can be used without appreciable error. Alternatively, the solution can be read directly from the Heisler chart.

Step 3
  Calculate the temperature:

  i) From the equation:
     The first root of the equation $\delta \tan \delta = Bi$ for $Bi = 1.162$ is: $\delta_1 = 0.9017$ (see Table A.2):

$$\frac{T_e - T_z}{T_e - T_o} \approx \frac{2Bi_z}{\delta_n^2 + Bi_z^2 + Bi_z} \frac{\cos\left(\frac{z}{L_z}\delta_n\right)}{\cos(\delta_n)} \exp\left(-\delta_n^2 Fo_z\right) =$$

$$= \frac{2 \times 1.162}{0.9017^2 + 1.162^2 + 1.162} \frac{\cos\left(\frac{0}{0.007} \times 0.9017\right)}{\cos(0.9017)}$$

$$\exp\left(-0.9017^2 \times 0.894\right) = 0.545$$

  ii) From the Heisler chart for an infinite slab (Fig. A.5):

  • Find the value of $Fo = 0.89$ on the x-axis.
  • Find the curve with $k/hL = 0.86$ (interpolate between curves 0.8 and 1.0).
  • Read the dimensionless temperature on the y-axis as:

$$\frac{T_e - T_z}{T_e - T_o} = 0.55$$

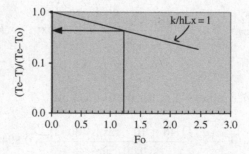

4) The combined effect of heat transferred in the x, y, and z directions is:

$$\frac{T_e - T_{xyz}}{T_e - T_o} = \left(\frac{T_e - T_x}{T_e - T_o}\right)\left(\frac{T_e - T_y}{T_e - T_o}\right)\left(\frac{T_e - T_z}{T_e - T_o}\right)$$
$$= 0.455 \times 0.618 \times 0.545 = 0.153$$

and the temperature is:

$$\frac{100 - T_{xyz}}{100 - 20} = 0.153 \rightarrow T_{xyz} = 87.8\,^\circ C$$

Example 10.4

Calculate how long it will take for the temperature on the non-heated surface of the steak of Example 10.1 to increase by 1 % of the initial temperature difference.

Solution

Step 1
Identify the geometry of the system.
As long as the temperature change on the cold surface of the steak is less than 1% of the initial temperature difference, the steak can be treated as a semi-infinite body.

Step 2
Examine the surface temperature.
The surface temperature is constant at 120°C.

Step 3
Select the appropriate equation.
The equation for a semi-infinite body with uniform initial temperature and constant surface temperature is (Table 10.4):

$$\frac{T_e - T}{T_e - T_o} = erf\left(\frac{x}{2\sqrt{\alpha t}}\right)$$

with

$$\frac{T - T_o}{T_e - T_o} = 0.01$$

(since the accomplished temperature change is 1%)

or

$$\frac{T_e - T}{T_e - T_o} = 1 - \frac{T - T_o}{T_e - T_o} = 1 - 0.01 = 0.99$$

Therefore

$$0.99 = \mathrm{erf}\left(\frac{x}{2\sqrt{\alpha t}}\right)$$

Step 4
Find the erf of the argument $x/2\sqrt{\alpha t}$ from an erf table (Table A.5).
The argument has to be equal to 1.82 for the error function to be equal to 0.99.
Therefore

$$\frac{x}{2\sqrt{\alpha t}} = 1.82$$

Step 5
Solve for t:

$$t = \frac{\left(\frac{x}{3.64}\right)^2}{\alpha} = \frac{\left(\frac{0.02 \text{ m}}{3.64}\right)^2}{1.4 \times 10^{-7} \, \text{m}^2/\text{s}} = 215.6 \text{ s}$$

Comment: The distance x where the temperature has increased by 1% of the initial temperature difference is called the "thermal penetration depth." As long as the thermal penetration depth in a finite body is less than the thickness of the body (or half the thickness of the body, in the case of heating from both sides), the body can be treated as a semi-infinite body.

Example 10.5

Concentrated milk is sterilized in a can with 7.5 cm diameter and 9.5 cm height. Calculate the time required to heat the milk from 45°C to 115°C given that: a) the can is in such a motion during heating that the milk temperature is uniform inside the can, b) the can contains 410 g of milk, c) the overall heat transfer coefficient between the heating medium and the milk is $300\text{W}/\text{m}^2°\text{C}$, d) the heating medium temperature is 130°C, and e) the mean heat capacity of the milk over the temperature range of the heating process is 3650J/kg°C.

Solution

Since the milk temperature is uniform inside the can during heating, the internal resistance can be considered negligible. The solution for $Bi < 0.1$ can be applied (Table 10.3). Thus,

$$t = \frac{mc_p}{hA} \ln \frac{T_e - T_o}{T_e - T} =$$

$$= \frac{(0.410 \, kg)(3650 \, J/kg^\circ C)}{\left(300 \, W/m^2 {}^\circ C\right) \left(\pi \, (0.075 \, m)(0.095 \, m) + 2 \left(\pi \, \frac{0.075^2}{4}\right) m^2\right)} \ln \frac{130 - 45}{130 - 115}$$

$$= 277 \, s$$

Exercises

Exercise 10.1

Peas are blanched by immersion in hot water at $90^\circ C$. Calculate the temperature at the center of a pea after 3 min if the diameter of the pea is 8 mm, the initial temperature of the pea is $20^\circ C$, and the heat transfer coefficient at the surface of the pea is $100 W/m^2 {}^\circ C$. Assume that the physical properties of the pea are $\rho = 1050 kg/m^3$, $c_p = 3.7 kJ/kg^\circ C$, $k = 0.5 W/m^\circ C$.

Solution

Step 1
State your assumptions.
The pea can be assumed to be a sphere.

Step 2
Select the equation for a sphere.

i) Calculate the Biot number:

$$Bi = \text{...}$$

Since $0.1 < Bi < 40$, internal and external resistances are important.

ii) The equation to be used is:

$$\frac{T_e - T}{T_e - T_o} = \sum_{n=1}^{\infty} \frac{2 Bi}{\delta_n^2 + Bi^2 - Bi} \frac{\delta_n}{\sin(\delta_n)} \exp\left(-\delta_n^2 Fo\right)$$

Step 3
Calculate the Fourier number:

$$\alpha = \frac{\text{..............}}{\text{..........................}} = \text{..............}$$

and

$$Fo = \text{...}$$

Since Fo > 0.2, only the first term of the sum in the above equation can be used without appreciable error. Alternatively, the solution can be read directly from the Heisler chart.

Step 4
Calculate the temperature:

i) From the above equation:
   Substitute values in the above equation ($\delta_1 = 1.432$) and calculate the temperature as

$$\frac{90 - T}{90 - 20} \approx \text{...}$$

$$T = \text{..}°C$$

ii) Also find the temperature from the Heisler chart for a sphere:

   • Find the value of Fo on the x-axis.
   • Calculate k/hR and find the corresponding curve.

$$\frac{k}{hR} = \text{.....................................}$$

   • Read the dimensionless temperature on the y-axis as:

$$\frac{T_e - T}{T_e - T_o} = \text{...........................}$$

   • Calculate the temperature as:

$$\frac{\text{.................... } - T}{\text{..................................}} = \text{.......}$$

$$\rightarrow T = \text{..........................}$$

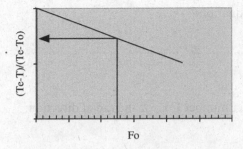

Exercise 10.2

A 211x300 can containing a meat product is heated in a retort. The initial temperature of the can is 50°C, the heat transfer coefficient at the surface is $3000 \text{W}/\text{m}^2\text{°C}$, and the physical properties of the meat product are: $k = 0.5\text{W}/\text{m°C}$, $c_p = 2.85\text{kJ}/\text{m°C}$, and $\rho = 1100\text{kg}/\text{m}^3$. Calculate the temperature at the geometric center of the can after 30 min if the steam temperature in the retort is 130°C.

Solution

Step 1
Determine the shape of the object.
The shape of the can is a finite cylinder. The temperature at the center is affected by the heat transferred from the cylindrical surface as well as from the flat bases of the cylinder. The contribution from each direction has to be calculated separately and the combined effect will be calculated at the end.

The dimensions of the can are:

$$\text{Diameter} = 2\frac{11}{16} \text{ inches} = 0.06826 \,\text{m},$$

$$\text{Height} = 3\frac{0}{16} \text{inches} = 0.0762 \,\text{m}$$

1) Calculate the heat transferred in the radial direction.

Step 1
Select the appropriate equation:

    i) Calculate the Biot number for the r-direction and determine the relative significance of external and internal resistances:

$$\text{Bi}_r = \dots\dots\dots\dots\dots\dots\dots\dots\dots\dots\dots\dots\dots\dots\dots$$

ii) Select the equation:

....

Step 2
Calculate the Fourier number Fo_r for the radial direction:

$$Fo_r = \text{...}$$

Is $Fo_r > 0.2$? If yes, only the first term of the sum in the above equation can be used without appreciable error. Alternatively, the solution can be read directly from the Heisler chart.

Step 3
Calculate the temperature:

i) From the equation:

Find the values of δn. As in Example 10.2, substitute values in the selected equation and calculate:

$$\frac{T_e - T_r}{T_e - T_o} = \text{...}$$

ii) From the Heisler chart:
On the x-axis of the Heisler chart for a infinite cylinder:

• Find the value of Fo_r.
• Calculate k/hR and find the corresponding curve.
• Read the dimensionless temperature on the y-axis as:

$$\frac{T_e - T_r}{T_e - T_o} = \text{...}$$

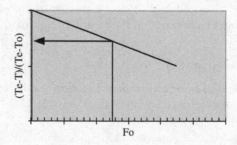

2) Calculate the heat transferred in the x-direction.

Step 1
Select the appropriate equation:

    i) Calculate the Biot number for the x-direction and determine the relative significance of external and internal resistances:

$$Bi_x = \text{...}$$

    ii) Select the equation:

....

Step 2
Calculate the Fourier number Fo_x for the x-direction:

$$Fo_x = \text{...}$$

Is $Fo_x > 0.2$? If yes, only the first term of the sum in the above equation can be used without appreciable error. Alternatively, the solution can be read directly from the Heisler chart.

Step 3
Calculate the temperature:

    i) From the equation:
    Substitute values and calculate:

$$\frac{T_e - T_x}{T_e - T_o} = \text{..}$$

    ii) From the Heisler chart:
    On the x-axis of the Heisler chart for an infinite slab:

- Find the value of Fo_x.
- Calculate k/hL_x and find the corresponding curve.
- Read the dimensionless temperature on the y-axis as:

$$\frac{T_e - T_x}{T_e - T_o} = \text{.........................}$$

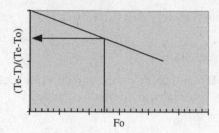

3) Calculate the combined effect of the heat transferred through the cylindrical surface and the heat transferred through the flat bases of the can:

$$\frac{T_e - T_{rx}}{T_e - T_o} = \left(\frac{T_e - T_r}{T_e - T_o}\right)\left(\frac{T_e - T_x}{T_e - T_o}\right) = \dots\dots\dots\dots\dots\dots\dots\dots\dots\dots\dots\dots\dots\dots$$

Calculate the temperature as:

$$\frac{130 - T_{rx}}{\dots\dots\dots\dots\dots\dots\dots} = \dots\dots\dots\dots\dots \rightarrow T_{rx} = \dots\dots\dots\dots\dots °C$$

Exercise 10.3

For the removal of field heat from fruits, hydrocooling is usually used. Calculate the time required to reduce the center temperature of a spherical fruit from 25°C to 5°C by immersion in cold water of 1°C if the diameter of the fruit is 6 cm, the heat transfer coefficient is $1000 W/m^2°C$, and the physical properties of the fruit are: $k = 0.4 W/m°C$, $\rho = 900 kg/m^3$, $c_p = 3.5 kJ/kg°C$.

Solution

Solve the problem using the Heisler chart for a sphere:

- Calculate the value of the dimensionless temperature:

$$\frac{T_e - T}{T_e - T_o} = \dots\dots\dots\dots\dots\dots\dots\dots\dots\dots$$

- Find the value of the dimensionless temperature on the y-axis of the Heisler chart:
- Calculate the value of k/hR and find the corresponding curve.

$$\frac{k}{hR} = \dots\dots\dots\dots\dots\dots\dots\dots\dots\dots$$

- Read the Fourier number on the x-axis (check if Fo > 0.2 so that the Heisler chart is valid).
- Calculate the time from:

$$t = \frac{FoR^2}{\alpha} = \dots\dots\dots\dots\dots\dots\dots$$

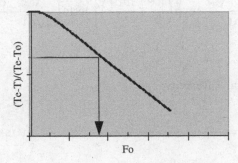

Exercise 10.4

A root crop is in the ground 5 cm from the surface when suddenly the air temperature drops to $-10°C$. If the freezing point of the crop is $-1°C$, calculate if the temperature of the soil at this depth will drop below the freezing point after 24 h. The initial temperature of the soil is $12°C$, the heat transfer coefficient at the surface of the ground is $10W/m^2\,°C$, and the physical properties of the soil are $k = 0.5W/m\,°C$, $\rho = 2000kg/m^3$ and $c_p = 1850J/kg\,°C$. Neglect latent heat effects.

Solution

Step 1
Draw the process diagram:

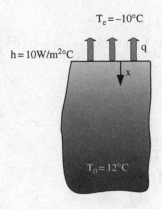

Step 2
Select the appropriate equation to use.The ground can be treated as a semi-infinite body. Therefore, the equation for unsteady state in a semi-infinite body with convection at the surface (since the surface temperature is not constant) will be used (Table 10.4):

$$\frac{T_e - T}{T_e - T_o} = erf\left(\frac{x}{2\sqrt{\alpha t}}\right) + \left[exp\left(\frac{hx}{k} + \frac{h^2 \alpha t}{k^2}\right)\right]\left[erfc\left(\frac{x}{2\sqrt{\alpha t}} + \frac{h\sqrt{\alpha t}}{k}\right)\right] \qquad (10.3)$$

Step 3
Calculate the thermal diffusivity:

$$\alpha = \frac{\dots\dots\dots}{\dots\dots\dots\dots\dots\dots\dots} = \dots\dots\dots\dots \; m^2/s$$

Step 4
Calculate:

$$\frac{x}{2\sqrt{\alpha t}} = \frac{0.05 \; m}{\dots\dots\dots\dots\dots\dots\dots\dots\dots\dots\dots} = \dots\dots\dots$$

and

$$\frac{h\sqrt{\alpha t}}{k} = \frac{\left(10 \; W/m^2 \, {}^\circ C\right) \times \dots\dots\dots\dots\dots\dots\dots\dots}{\dots\dots\dots\dots\dots\dots\dots\dots\dots\dots\dots} = \dots\dots\dots$$

Step 5
Find the value of $erf\left(\frac{x}{2\sqrt{\alpha t}}\right)$ and $erfc\left(\frac{x}{2\sqrt{\alpha t}} + \frac{h\sqrt{\alpha t}}{k}\right)$ from Table A.5 taking into account that erfc $= 1 - erf$.

Step 6
Substitute values in eqn (10.3) above and calculate:

$$\frac{T_e - T}{T_e - T_o} = \dots\dots\dots\dots\dots\dots$$

$$\rightarrow T = \dots\dots\dots\dots\dots {}^\circ C$$

Exercise 10.5

Calculate the temperature of 200 litres of glucose syrup heated in an agitated jacketed vessel for 30 min if the initial temperature of the syrup is 30°C, the temperature of the steam in the jacket is 120°C, the density and the heat capacity of the syrup are 1230kg/m³ and 3.35kJ/kg °C respectively, the overall heat transfer coefficient between the heating medium and the syrup is 200W/m² °C, and the inside surface area of the vessel in contact with the syrup is 1.5 m².

Solution

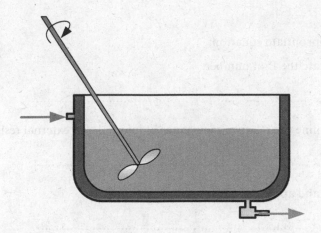

Step 1
Select the appropriate equation to use.Since the syrup is agitated, its temperature can be assumed to be uniform. Therefore the equation to use is:

....

Step 2
Calculate the mass of the syrup:

$$m = V\rho = \text{...}$$

Step 3
Apply the selected equation:

$$\frac{T_e - T}{T_e - T_o} = \text{...}$$

$$\rightarrow T = \text{........................}°C$$

Exercise 10.6

A thermocouple with 4 mm diameter and 2 cm length is immersed in milk which is at 60°C. How long it will take for the thermocouple to reach a temperature of 59.5°C if the heat transfer coefficient at the surface of the thermocouple is 200W/m² °C, the initial temperature of the thermocouple is 25°C, its mass is 1 g, its heat capacity is 0.461kJ/kg°C, and its thermal conductivity is 15W/m °C. If the thermocouple were exposed to air of 60°C instead of milk, how long would it take to reach 59.5°C? Assume that the heat transfer coefficient in this case is 20W/m² °C.

Solution

Step 1
Select the appropriate equation:

   i) Calculate the Biot number:

   Bi = ...

   ii) Determine the relative importance of internal and external resistance:

   ....

   iii) Select and apply the appropriate equation:

   ....

Step 2
For the case of air, repeat the calculations with h = 20W/m²°C.

Exercise 10.7

Solve Example 10.1 using the spreadsheet program *Heat Transfer-Negligible Surface Resistance.xls*. Also solve the problem for the case in which the thickness of the steak is 3 cm and for the case in which the steak is heated from both sides.

Solution

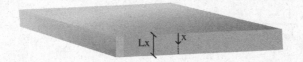

Step 1
On the spreadsheet *Heat Transfer-Negligible Surface Resistance.xls,* go to the sheet "slab.
"Turn the "SWITCH" OFF (set the value in cell G1 equal to 0 and press ENTER).

Step 2
Insert the parameter values in the yellow cells using: half thickness in the x-direction L_x = 0.02 m (since the steak is heated from one side, L_x is equal to the thickness); half thickness in the y-direction and z-direction L_y = 1 m, L_z = 1 m (since the steak is considered as an infinite plate, use big numbers on L_y and L_z to simulate the fact that heat transfer is significant only in the x-direction); distance of the point from the coldest plane in the x-direction x = 0.02 − 0.011 = 0.009m; distance in the z and y directions y = 0 and z = 0 (since the steak is considered as an infinite plate); initial temperature T_o = 5°C; surface temperature T_e = 120°C. Also insert the given values for the thermal diffusivity.

Step 3
Set the value of the time step to 15 × 60 = 900 s in cell F4.

Step 4
Turn the "SWITCH" ON (set the value in cell G1 equal to 1 and press ENTER).

Step 5
Read the value for the temperature of the point that lies 9 mm below the surface in cell K20 (green cell).

Step 6
Read the value for the mean temperature in cell K40 (green cell).

Step 7
To see the effect of steak thickness, set L_x = 0.03. Repeat steps 2 to 6.

Step 8
To see how the temperature of the point changes with time:

a) Turn the "SWITCH" OFF (set the value in cell G1 equal to 0 and press ENTER).
b) Set the value of the time step in cell F4 equal to 10 or some other small value.
c) Turn the "SWITCH" ON (set the value in cell G1 equal to 1 and press ENTER).
d) Iterate by pressing F9 until the temperature in cell K20 reaches the value you want.

Step 9
To see how much faster the temperature at the same point would have increased
if the steak were heated from both sides, i.e., between the hot plates of a toaster:

a) Turn the "SWITCH" OFF (set the value in cell G1 equal to 0 and press
   ENTER).
b) Set the value of half thickness in the x-direction to L_x = 0.01 m.
c) Set the value of the coordinates of the point to x = 0.001 m, y = 0, z = 0.
d) Set the value of the time step equal to 10.
e) Turn the "SWITCH" ON (set the value in cell G1 equal to 1 and press
   ENTER).
f) Iterate by pressing F9 until the same temperatures as in steps 5 and 6 are
   reached.
g) Read the time in cell F3.

Exercise 10.8

Solve Example 10.2 using the spreadsheet program *Heat Transfer-Negligible
Surface Resistance.xls*.

Solution

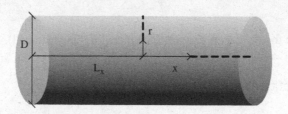

Step 1
On the spreadsheet *Heat Transfer-Negligible Surface Resistance.xls,* go to the
sheet "cylinder." Turn the "SWITCH" OFF (set the value in cell G1 equal to 0
and press ENTER).

Step 2
Insert the parameter values in the yellow cells using: diameter of the cylinder
D = 0.015 m; half of the height of the cylinder L_x = 0.08 m; distance of
the point from the center line in the direction of radius r = 0.0045 m

(point 3 mm from the surface); distance of the point from the center of the cylinder in the x direction $x = 0$; initial temperature $T_o = 5°C$; surface temperature $T_e = 100°C$. Also insert the given values for the physical properties and the heat transfer coefficient.

Step 3
Read the Biot number for the r and x directions in cells B42 and C42

Step 4
Set the value of the time step to 120 in cell F4.

Step 5
Turn the "SWITCH" ON (set the value in cell G1 equal to 1 and press ENTER).

Step 6
Read the value for the temperature of the point that lies 3 mm below the surface in cell K20 (green cell).

Step 7
Turn the "SWITCH" OFF (set the value in cell G1 equal to 0 and press ENTER).

Step 8
Set the value of $r = 0$ and $x = 0$ (coordinates of the center of the cylinder).

Step 9
Turn the "SWITCH" ON (set the value in cell G1 equal to 1 and press ENTER).

Step 10
Read the value for the temperature of the center of the hot dog in cell K20 (green cell).

Step 11
Turn the "SWITCH" OFF (set the value in cell G1 equal to 0 and press ENTER).

Step 12
Set the value of the time step to 0.5 in cell F4.

Step 13
Turn the "SWITCH" ON (set the value in cell G1 equal to 1 and press ENTER).

Step 14
Iterate by pressing F9 until the temperature in cell K20 is equal to 81°C. Read the value of the time in cell F3.

Step 15
Turn the "SWITCH" OFF (set the value in cell G1 equal to 0 and press ENTER).

Step 16
Set the value of the time step to 120 in cell F4.

Step 17
Turn the "SWITCH" ON (set the value in cell G1 equal to 1 and press ENTER).

Step 18
Read the value for the mean temperature in cell K40 (green cell).

Step 19
Read the value for the heat transferred in cell K43.

Exercise 10.9

Solve Example 10.3 using the spreadsheet *Heat Transfer-Internal and External Resistance.xls* . Run the program for h = 50 W/m² °C and h = 200 W/m² °C to see the effect of the heat transfer coefficient on the time-temperature relationship.

Solution

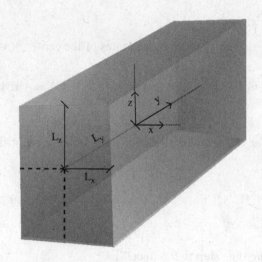

Step 1
On the spreadsheet *Heat Transfer-Internal and External Resistance.xls* go to the sheet "slab".
Turn the "SWITCH" OFF (set the value in cell G1 equal to 0 and press ENTER).

Step 2
Insert the parameter values in the yellow cells using: $L_x = 0.006$ m; $L_y = 0.008$ m; $L_z = 0.007$ m; distance of the point from the geometric center x = 0, y = 0, z = 0; initial temperature $T_o = 20°C$; temperature of the environment $T_e = 100°C$.

Also insert the given values for the density, the heat capacity, the thermal conductivity, and the heat transfer coefficient.

Step 3
Read the Biot numbers for the x, y, and z direction in cells B37, C37 and D37. Are they between 0.1 and 40?

Step 4
Set the value of the time step to $5 \times 60 = 300$ in cell F4.

Step 5
Turn the "SWITCH" ON (set the value in cell G1 equal to 1 and press ENTER).

Step 6
Read the temperature for the center in cell K15 (green cell).

Step 7
Read the mass average temperature in cell K30 (green cell).

Step 8
To see how the temperature at the selected point changes with time turn the "SWITCH" OFF (set the value in cell G1 equal to 0 and press ENTER).

Step 9
Set the value of the time step equal to 10 (or some other small value) in cell F4.

Step 10
Turn the "SWITCH" ON (set the value in cell G1 equal to 1 and press ENTER).

Step 11
Iterate (by pressing F9) until the value of time in cell F3 (grey cell) reaches 300.

Step 12
See the plot for the center temperature and mass average temperature vs. time and the dimensionless temperature in the x-direction vs. Fourier number for the x-direction in the diagrams.

Step 13
Change the value of h and the corresponding values of δi. Run the program and see the temperature at the center after 5 min.

Step 14
Make the necessary changes in the spreadsheet to plot the dimensionless non-accomplished temperature change for the y and z directions.

Exercise 10.10

For the removal of field heat from certain fruits, hydrocooling, forced air cooling and room cooling have been proposed. Calculate and plot the temperature change at the center of the fruit vs. the time for the three methods. Assume that the fruit is spherical; its initial temperature is 25°C; the cold water and the air temperature are 1°C; the diameter of the fruit is 6 cm; the heat transfer coefficient is $1000 \text{W}/\text{m}^2 {}°\text{C}$ in the case of hydrocooling, $20 \text{W}/\text{m}^2 {}°\text{C}$ in the case of forced air cooling, and $5\text{W}/\text{m}^2 {}°\text{C}$ in the case of room cooling; and the physical properties of the fruit are: $k = 0.4\text{W}/\text{m}°\text{C}$, $\rho = 900\text{kg}/\text{m}^3$, $c_p = 3.5\text{kJ}/\text{kg}°\text{C}$.

Solution

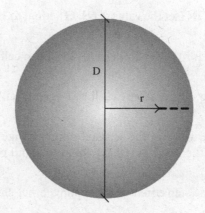

Step 1
Calculate the Biot number in all cases.

Step 2
Select the spreadsheet *Heat Transfer-Negligible Surface Resistance.xls* or *Heat Transfer- Internal and External Resistance.xls* depending on the value of Biot number.

Step 3
Follow the instructions and run the program for a sphere.

Step 4
Run the program using a small time increment, e.g., 10 s.

Exercise 10.11

The product development department in a food plant is interested in knowing what shape they should give to a new product for a faster temperature change at the center of the product from 20°C to 90°C. Should the shape be a cube 50mm × 50mm × 50mm, a cylinder with diameter 62.035 mm and length 41.36 mm, or a sphere with diameter 62.035 mm? All these shapes will have the same volume and the same mass. The heating medium is steam at 100°C; the heat transfer coefficient in all cases is $2000W/m^2°C$; and the density, the heat capacity, and the thermal conductivity of the product are $900kg/m^3$, $3200J/kg°C$, and $0.5W/m°C$ respectively. Determine which shape will give a faster temperature change at the center.

Solution

Step 1
Calculate the Biot number in all cases.

Step 2
Select the spreadsheet *Heat Transfer-Negligible Surface Resistance.xls* or *Heat Transfer- Internal and External Resistance.xls* depending on the value of the Biot number.

Step 3
Follow the instructions and run the program for a cube, a cylinder, and a sphere.

Step 4
Find the time necessary in each case for T_{local} to reach 90°C.

Exercise 10.12

Understand how the spreadsheet program *Heat Transfer-Negligible Surface Resistance.xls* calculates and plots the temperature change of a sphere with time. Modify the program to plot the temperature distribution with the radius of the sphere for a certain time.

Exercise 10.13

Understand how the spreadsheet program *Heat Transfer-Internal and External Resistance.xls* calculates and plots the temperature change of a cylinder with time. Modify the program to plot q_t/q_evs.$Fo^{0.5}$ for a certain Biot number value for an infinite cylinder.

Exercise 10.14

a) Solve Exercise 10.4 using the spreadsheet program *Semi-Infinite body 1.xls*.
b) Plot the temperature profile for the depth 0-5 cm after 12 h and 24 h. c) What is the minimum depth at which the temperature does not change after 12 h?
d) Observe how the temperature changes with time at the depth of 1 cm, 3 cm, and 5 cm for the first 24 hours.

Solution

Step 1
In the sheet "T vs. x", follow the instructions, run the program for Time = $24 \times 3600 = 86400$ s and read the temperature in cell F16 when the distance in cell F3 is equal to 0.05 m. Observe the temperature profile in the plot.

Step 2
Set the time equal to $12 \times 3600 = 43200$ s in cell B29, follow the instructions, run the program, and observe the temperature profile in the plot.

Step 3
Continue running the program until the temperature in cell F16 is equal to 12°C. Read the depth in cell F3.

Step 4
In the sheet "T vs. t", follow the instructions, run the program for Distance x = 0.01 m until Time = 86400 s in cell F3. Observe the temperature profile in the plot. Rerun the program for x = 0.03 m and x = 0.05 m following the instructions.

Exercise 10.15
Use the spreadsheet program *Semi-Infinite body 2.xls*, to study how the temperature in a semi-infinite body changes with time when a) the temperature at the surface is constant, and b) the temperature at the surface is not constant but is affected by convection. Find the value of $h\sqrt{\alpha t}/k$ for which the solution for both cases is approximately the same.

Solution

Step 1
In the sheet "Convection at the surface", follow the instructions and run the program for various values of $h\sqrt{\alpha t}/k$ in cell B41.

Step 2
Read the temperature in cell H16.

Step 3
For the case in which the temperature at the surface is constant, run the sheet "Constant surface temperature."

Step 4
Observe the temperature change in the Temperature vs. Time plot. Observe the difference between the two solutions for low values of $h\sqrt{\alpha t}/k$. Which is the correct one?

Step 5
Observe the plot of accomplished temperature change $\frac{T-T_0}{T_e-T_0}$ vs. $\frac{x}{2\sqrt{\alpha t}}$.
Compare it with the respective plot in your textbook, e.g., Ref 3, 4, 5.

Exercise 10.16

Using the spreadsheet program *Semi-Infinite body 3.xls,* calculate the time necessary for the thermal penetration depth to reach the non-heated surface of the steak of Example 10.1. If the temperature of the hotplate changes to 150°C, does the time necessary for the thermal penetration depth to reach the non-heated surface of the steak change? Is the temperature at x = 0.02 m the same as in the case of 120°C at the time the thermal penetration depth is equal to the thickness of the steak?

Solution

Step 1
In the sheet "Calculations," follow the instructions and insert the parameter values. For the heat transfer coefficient use a high value, e.g., 3000W/m²°C in cell B39 (assume a negligible conduct resistance).

Step 2
Run the program until Time in cell F4 remains constant. Read the required time in cell G16.

Step 3
Observe the temperature change in the Temperature vs. Distance plot.

Step 4
Use a hotplate temperature in cell B31 of 150°C. Rerun the program until Time in cell G16 remains constant. Read the required time in cell G16.

Step 5
Read the temperature in cell G18 when x = 0.02 in cell G17. Compare this value with the corresponding temperature when the hot plate temperature is 120°C.

Exercise 10.17

In the "hot-break" process, whole tomatoes are heated with live steam to inactivate the pectin methylesterase enzyme before crushing and extraction, thereby yielding high viscosity products. Calculate the time necessary to heat the center of a tomato 6 cm in diameter to 50°C using 100°C steam if the initial temperature of the tomato is 20°C and the density, the heat capacity, and the thermal conductivity for tomatoes are 1130kg/m³, 3.98kJ/kg°C, and 0.5W/m°C respectively. Find the time necessary for the mass average temperature to reach a value of 82 °C.

Solution

Use the spreadsheet *Heat Transfer-Negligible Surface Resistance.xls* to solve the problem. Write the necessary assumptions.

Exercise 10.18

A potato 5 cm in diameter is immersed in boiling water. Calculate the required time for the temperature at the center of the potato to reach 59°C, the onset of starch gelatinization, if the initial temperature of the potato is 21°C and the density, the heat capacity and the thermal conductivity for potatoes are 1120kg/m^3, $3.7 \text{kJ/kg}\,°C$ and $0.55 \text{W/m}\,°C$ respectively. Find the mass average temperature of the potato when the center temperature is 59°C. Study the effect of diameter on the required time. Is it proportional to D or to D^2?

Solution

Use the spreadsheet *Heat Transfer-Negligible Surface Resistance.xls* to solve the problem. Write the necessary assumptions.

Chapter 11
Mass Transfer by Diffusion

Theory

As with heat transfer, the rate of the transferred quantity in mass transfer is proportional to the driving force and inversely proportional to the resistance. For mass transfer by diffusion, the driving force is the chemical potential difference; but concentration (mol/m^3), mole fraction, or pressure differences are usually used. Thus, the mass transfer rate is:

$$n_i = \frac{\text{driving force}}{\text{resistance}} = \frac{\Delta c_i}{R}$$

For gases:

	Driving force	Resistance, R
Equimolar Counterdiffusion	Δp_i	$\dfrac{\Delta z}{D_{ij}A} R_G T$
Diffusion of i through stagnant j	Δp_i	$\dfrac{\Delta z}{D_{ij}A} \dfrac{p_{jM}}{P} R_G T$

For liquids:

	Driving force	Resistance, R
Equimolar Counterdiffusion	Δx_i	$\dfrac{\Delta z}{D_{ij}A} \dfrac{1}{c_m}$
Diffusion of i through stagnant j	Δx_i	$\dfrac{\Delta z}{D_{ij}A} \dfrac{x_{jM}}{c_m}$

For solids:

	Mass transfer rate
For a single wall	$n_i = \dfrac{\Delta c_i}{R}$
For a composite wall	$n_i = \dfrac{\Delta c_i}{\sum R}$

	Resistance, R	Surface area, A
Flat wall	$\dfrac{\Delta z}{DA}$	A
Cylindrical wall	$\dfrac{\Delta r}{DA_{LM}}$	$A_{LM} = \dfrac{A_1 - A_2}{\ln A_1/A_2}$
Spherical wall	$\dfrac{\Delta r}{DA_G}$	$A_G = \sqrt{A_1 A_2}$

Flat packaging films:

	Resistance, R	Surface area, A
For gases	$22414 \dfrac{\Delta z}{P_M A}$	A
For water vapor	$\dfrac{\Delta z}{P_{M\,WV} \cdot A}$	A

where

A = surface area perpendicular to the direction of transfer, m^2
A_{LM} = logarithmic mean of surface area A_1 and A_2, m^2
A_G = geometric mean of surface area A_1 and A_2, m^2
c_i = concentration of i, mol/m^3
c_m = mean concentration of i and j, mol/m^3
D = diffusion coefficient, m^2/s
D_{ij} = diffusion coefficient of i in j, m^2/s
n_i = mass transfer flux, rate, mol/s
P = total pressure, Pa
p_i = partial pressure of i, Pa
p_{jM} = logarithmic mean pressure difference of j, Pa
P_M = gas permeability, cm^3 cm/s cm^2 atm
$P_{M\,wv}$ = water vapor permeability, g cm/s cm^2 Pa
R_G = ideal gas constant, m^3 Pa/mol K
T = temperature, K

x_i = mol fraction of i
x_{jM} = logarithmic mean mol fraction of j
Δz and Δr = wall thickness, m

Review Questions

Which of the following statements are true and which are false?

1. Fick's 1st law refers to mass transfer by diffusion at steady state.
2. Molecular diffusion is a phenomenon analogous to heat transfer by conduction.
3. The resistance to mass transfer increases linearly with diffusivity.
4. Mass diffusivity has the same units as thermal diffusivity and kinematic viscosity.
5. Diffusivity in gasses is about 10000 times higher than in liquids.
6. Diffusivity does not vary with temperature.
7. Diffusivity in solids may vary with concentration.
8. The driving force for mass transfer by molecular diffusion is the difference in chemical potential.
9. There is bulk flow in equimolar counterdiffusion.
10. In equimolar counterdiffusion $N_i = -N_j$, where N_i and N_j are the fluxes of gas i and gas j with respect to a fixed position.
11. There is no bulk flow in the case of diffusion of gas i through stagnant nondiffusing gas j.
12. For the same driving force, the flux N_i in equimolar counterdiffusion is smaller than N_i in diffusion of i through stagnant nondiffusing j due to the bulk motion of i.
13. When concentrations are dilute, the bulk flow may be negligible.
14. Permeability refers to the diffusion of a gas in a solid and is used extensively in calculating mass transfer in packaging materials.
15. Permeability is equal to the product of the diffusion coefficient and the solubility of the gas in the solid.
16. The difference in partial pressure inside and outside the packaging material is being used as the driving force for mass transfer calculations.
17. Permeability decreases as the temperature increases.
18. Layers of different materials may be combined in laminates to give a composite material with good barrier properties for water vapor, gasses, and light.
19. Polyethylene is a good water vapor barrier and serves as an adhesive to the next layer.
20. Aluminum foil is a good gas barrier

Examples

Example 11.1

Water vapor is diffusing through a stagnant film of air at 30°C towards a cold surface at 6°C, where it is condensed. Calculate the water vapor diffusion flux if the pressure is 1 atm, the water vapor pressure 10 mm from the cold surface is 3 kPa, and the water vapor diffusion coefficient in the air is 0.26 cm^2/s.

Solution

Step 1
Draw the process diagram:

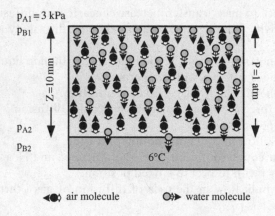

Step 2
State your assumptions:

- The system is at steady state.
- There are no eddies.

Step 3
Select the appropriate equation to calculate the water vapor diffusion flux.
Water molecules diffuse towards the cold plate; air molecules diffuse in the opposite direction. There is a bulk movement towards the cold plate to keep the system at constant pressure. Therefore the equation to use is:

$$N_w = \frac{n_w}{A} = \frac{p_{w1} - p_{w2}}{R\,A} = \frac{p_{w1} - p_{w2}}{\frac{\Delta z}{D_{aw}A}\frac{p_{aM}}{P}R_G\,T\,A} = \frac{D_{aw}P}{R_G\,T\,\Delta z\,p_{aM}}(p_{w1} - p_{w2})$$

(with subscript "w" for water vapor and "a" for air).

Step 4

Find the values of partial pressure to use in the above equation:

i) The water vapor partial pressure at the interphase at 6°C is

$$p_{w2} = 0.935 \text{ kPa(from steam tables)}.$$

ii) The partial pressure of air is

$$p_{a1} = 101325 - 3000 = 98325 \text{ Pa}$$

and

$$p_{a2} = 101325 - 935 = 100390 \text{ Pa}$$

$$p_{aM} = \frac{p_{a1} - p_{a2}}{\ln \dfrac{p_{a1}}{p_{a2}}} = \frac{98325 - 100390}{\ln \dfrac{98325}{100390}} = 99353.9 \text{ Pa}$$

Comment: The arithmetic mean instead of the log mean could have been used with very little error, since p_{a1} and p_{a2} values differ from one another by a small percentage.

Step 5

Substitute values and calculate the water vapor diffusion flux:

$$N_w = \frac{D_{aw}P}{RTz p_{aM}}(p_{w1} - p_{w2}) =$$

$$= \frac{(0.26 x 10^{-4} \text{ m}^2/\text{s})(101325 \text{ Pa})}{(8314.34 \text{ m}^3\text{Pa}/\text{kmol K})(303 \text{ K})(0.01 \text{ m})(99353.9 \text{ Pa})}(3000 - 935)\text{Pa} =$$

$$= 2.17 x 10^{-6} \text{ kmol/s m}^2$$

Example 11.2

A food product is sealed in a flexible laminated package with 0.1 m² surface area that is made of a polyethylene film layer 0.1 mm thick and a polyamide film layer 0.1 mm thick. The package is stored at 21°C and 75% relative humidity. Calculate the transfer rate of oxygen and water vapor through the film at steady state if the partial pressure of O_2 inside the package is 0.01 atm, that outside the package is 0.21 atm, and the water activity of the product inside the package is 0.3. The permeability (P_M) of polyethylene and polyamide to O_2 are 2280×10^{-11} and 5×10^{-11} cm^3/ (s cm^2 atm /cm) respectively, and the water vapor transmission rate (WVTR) for these materials measured at 37.8°C using 90% RH water vapor source and 0% RH desiccant sink are 6×10^{-11} and 37×10^{-11} g/(s cm^2 /cm) respectively.

Solution

Step 1
State your assumptions:

- The convective resistance to mass transfer on the two sides of the package are negligible compared to the diffusion resistance of the film.
- The WVTR at 21°C does not differ appreciably from that at 37.8°C.

Step 2
Calculate the diffusion rate of oxygen:

i) Select the equation to use:

$$n_{O_2} = \frac{\Delta p_{O_2}}{\sum R}$$

ii) Substitute values and calculate the diffusion rate:

$$\sum R = R_{\text{polyethylene}} + R_{\text{polyamid}} =$$

$$= \frac{22414 \, \Delta z_{\text{polyeth.}}}{A \, P_{M \text{polyeth.}}} + \frac{22414 \, \Delta z_{\text{polyam.}}}{A \, P_{M \text{polyam.}}} = \frac{22414}{A} \left(\frac{\Delta z_{\text{polyeth.}}}{P_{M \text{polyeth.}}} + \frac{\Delta z_{\text{polyam.}}}{P_{M \text{polyam.}}} \right) =$$

$$= \frac{(22414 \, \text{cm}^3/\text{mol})}{(1000 \, \text{cm}^2)} \left(\frac{(0.01 \, \text{cm})}{(2280 \times 10^{-11} \text{cm}^3 \, \text{cm/s cm}^2 \, \text{atm})} \right.$$

$$\left. + \frac{(0.01 \, \text{cm})}{(5 \times 10^{-11} \text{cm}^3 \, \text{cm/s cm}^2 \, \text{atm})} \right) =$$

$$= 9.83 \times 10^6 + 4.48 \times 10^9 = 4.49 \times 10^9 \, \text{atm s/mol}$$

$$n_{O_2} = \frac{\Delta p_{O_2}}{\sum R} = \frac{0.21 - 0.01 \, \text{atm}}{4.49 \times 10^9 \, \text{atm s/mol}} = 4.45 \times 10^{-11} \, \text{mol/s}$$

Step 3
Calculate the diffusion rate of water vapor.

i) Select the equation to use:

$$n_w = \frac{\Delta p_w}{\sum R}$$

with p_w the water vapor difference between the inside and the outside of the package and

$$R = \frac{\Delta z}{P_{MWV} A}$$

ii) Find the water vapor pressure inside and outside the package.
The water vapor pressure p_w at 21°C is 2487 Pa (from steam tables). The water vapor partial pressure outside the package for 75% relative humidity is:

$$p_{wo} = \frac{75}{100} \, 2487 = 1865.3 \, Pa$$

The water vapor partial pressure inside the package for water activity 0.3 is:

$$p_{wi} = 0.3 \times 2487 = 746.1 \, Pa$$

iii) Calculate the water vapor permeability.
Water vapor permeability for polyethylene can be calculated from the Water Vapor Transmission Rate as:

$$P_{MWV \, polyethylene} = \frac{WVTR_{polyeth.}}{\Delta p} = \frac{6 \times 10^{-11} g \, cm/s \, cm^2}{0.90 \times 6586 - 0 \, Pa}$$
$$= 1.01 \times 10^{-14} g \, cm/s \, cm^2 \, Pa$$

Similarly for polyamide:

$$P_{MWV \, polyamid} = \frac{WVTR_{polyeth.}}{\Delta P} = \frac{37 \times 10^{-11} g \, cm/s \, cm^2}{0.90 \times 6586 - 0 \, Pa}$$
$$= 6.24 \times 10^{-14} g \, cm/s \, cm^2 \, Pa$$

iv) Calculate the total resistance to water vapor transfer for the laminate:

$$\sum R = R_{polyet\,h\,y\,l\,e\,ne} + R_{polyamid} = \frac{\Delta z_{polyeth.}}{P_{M\,WV\,polyeth.}A} + \frac{\Delta z_{polyamid}}{P_{M\,wv\,polyamid}\,A} =$$

$$= \frac{0.01\ cm}{(1.01 \times 10^{-14} g\,cm/s\ cm^2\,Pa)\ (1000\ cm^2)}$$

$$+ \frac{0.01\ cm}{(6.24 \times 10^{-14} g\,cm/s\ cm^2\,Pa)\ (1000\ cm^2)} =$$

$$= 9.90 \times 10^8 + 1.60 \times 10^8 = 1.15 \times 10^9\ Pa\ s/g$$

v) Calculate the water vapor transfer rate:

$$n_w = \frac{\Delta p_w}{\sum R} = \frac{1865.3 - 746.1\ Pa}{1.15 \times 10^9\ Pa\ s/g} = 9.73 \times 10^{-7}\ g/s$$

Comment: Notice that the polyethylene film layer contributes the main resistance to water vapor transfer (86.1%), while polyamide contributes the main resistance to oxygen transfer (99.8%).

Exercises

Exercise 11.1

Water evaporates from the flat surface of the leaf of a vegetable and is diffusing away through a stagnant layer of air. The total pressure is 101325 Pa and the temperature is 24°C. Calculate the evaporation rate in g/s under the following conditions: water activity (a_w) of the leaf surface is 0.98, partial water vapor pressure 5 mm away from the surface of the leaf is 2100 Pa, surface area of the leaf is 50 cm². The diffusion coefficient of water vapor in air is 2.6×10^{-5} m²/s.

Solution

Step 1
State your assumptions:

The time interval the calculations are based on is small so that the water vapor pressure at the surface of the leaf is constant and the system is at steady state.

Step 2
Select the equation to use:

Since water vapor diffuses through a stagnant layer of air, the equation to be used is:

$$N_w = \dots\dots\dots\dots\dots\dots\dots\dots\dots\dots\dots\dots\dots\dots\dots\dots$$

Step 3
Find the water vapor pressure p_w at 24°C from the steam tables: ....
Calculate the partial water vapor pressure p_{w1} at the surface of the leaf as:

$$p_{w1} = a_w \cdot p_w = \dots\dots\dots\dots\dots\dots\dots\dots\dots\dots\dots$$

Step 4
Calculate the partial pressure of air:

$$p_{a1} = P - p_{w1} = \dots\dots\dots\dots\dots\dots\dots\dots\dots\dots\dots\dots\dots$$

$$p_{a2} = \dots\dots\dots\dots\dots\dots\dots\dots\dots\dots\dots\dots\dots\dots\dots$$

$$p_{aM} = \dots\dots\dots\dots\dots\dots\dots\dots\dots\dots\dots\dots\dots\dots$$

Step 5
Calculate the mass transfer flux:

$$N_w = \dots\dots\dots\dots\dots\dots\dots\dots\dots\dots\dots\dots\dots\dots\dots\dots\dots\dots\dots\dots\dots\dots\dots\dots.kmol/s\ m^2$$

Step 6
Calculate the evaporation rate:

$$n_w = \dots\dots\dots\dots\dots\dots\dots\dots\dots\dots\dots\dots\dots\dots\dots\dots\dots\dots\dots\dots\dots\dots\dots\dots\dots\dots.g/s$$

Exercise 11.2

An aroma compound is encapsulated in a nonporous spherical particle made of a homogenous biopolymer film. Calculate the aroma release rate if the particle diameter is 1 mm, the film thickness is 0.1 mm, the concentration of the aroma compound is $0.1\ g/cm^3$ and $0.01 g/cm^3$ on the inside and outside surface of the particle respectively, and the diffusion coefficient of the aroma compound in the film is $1 \times 10^{-12}\ m^2/s$.

Solution

Step 1
Draw the process diagram:

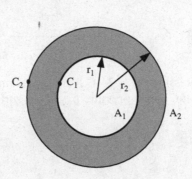

Step 2
State your assumptions:

- The aroma release rate is diffusion-controlled.
- The release rate is constant (steady state).

Step 3
Calculate the resistance to mass transfer in a spherical wall.

   i) Calculate the geometric mean area of A_1 and A_2:

$$A_1 = \pi D^2 = \text{..}$$

$$A_2 = \text{..}$$

$$A_G = \text{..}$$

   ii) Calculate the resistance:

$$R = \frac{\Delta r}{D A_G} \text{..}$$

Step 4
Calculate the diffusion rate through the wall of the sphere:

$$n = \text{...}$$

Exercise 11.3

The solubility and the diffusivity of O_2 in polyvinyl chloride are 3.84×10^{-4} cm^3 O_2 at STP/cmHg.cm^3 solid and 1.18×10^{-12} m^2/s respectively. Calculate the permeability of polyvinyl chloride to O_2.

Solution

The permeability is the product of solubility and diffusivity. Therefore:

$$P_M = S.D = \text{...} \frac{cm^3 \; cm}{s \; cm^2 Pa}$$

Exercise 11.4

A supplier of frozen hamburger patties is thinking of packaging them in 0.1 mm thick polypropylene film. The permeability of polypropylene to O_2 and water vapor is 1×10^{-8} cm^3/(s cm^2 atm /cm) and 8×10^{-15} g/(s cm^2 Pa/cm) respectively. The partial pressure of O_2 is 0 atm and that of water vapor is 0.12 Pa inside the package, while outside the package the partial pressure of O_2 is 0.21 atm and that of water vapor is 0.04 Pa. Calculate how much oxygen and how much water vapor will pass through the polypropylene film of 600 cm^2 surface area in one week.

Solution

Step 1
Sate your assumptions:

- The convective resistance to mass transfer on the two sides of the package is negligible compared to the diffusion resistance of the film.
- ....

Step 2
Calculate the transfer rate of O_2.

i) Calculate the resistance:

R = ...

ii) Calculate the transfer rate of oxygen:

n_{O_2} = ...mol/s

Step 3
Calculate the total amount of O_2 that will diffuse into the package in one week:

m_{O_2} = ... g

Step 4
Calculate the transfer rate of water vapor.

  i) Calculate the resistance:

$$R = \text{..}$$

  ii) Calculate the transfer rate of water vapor:

$$n_w = \text{...} \text{g/s}$$

Step 5
Calculate the total amount of water vapor that will diffuse out of the package in one week:

$$m_w = \text{...} \text{ g}$$

Exercise 11.5

Meat is packaged in a plastic disk covered with PVC film. Calculate the required thickness of the film so that 1×10^{-4} mol of oxygen enters the package through the film in 24 hours. The permeability of PVC to oxygen is 456×10^{-11} cm^3/ (s cm^2 atm /cm), the surface area of the film is 300 cm^2, and the partial pressure of oxygen inside and outside the package is 0 atm and 0.21 atm respectively.

Solution

Step 1
State your assumptions:
....

Step 2
Calculate the mass transfer rate in mol/s:

$$n_{O_2} = \text{..}$$

Step 3
Calculate the required resistance:

$$R = \text{...}$$

Step 4
Calculate the required thickness of the film

$$\Delta z = \frac{P_M\,A}{22414} R = \text{...}$$

Exercise 11.6

Potato chips are packed in a plastic bag in a nitrogen atmosphere. Calculate the amount of N_2 that will diffuse out of the plastic bag in 3 months if the thickness of the film is 0.05 mm, the absolute pressure of N_2 in the bag is 1.05 atm, the surface area of the bag is 1000 cm^2, and the permeability of the plastic film to N_2 is 1.5×10^{-10} cm^3/s cm^2 atm/cm.

Solution

Step 1
State your assumptions:
....
Step 2
Calculate the resistance:

$$R = \text{...}$$

Step 3
Calculate the transfer rate of N_2:

$$n_{N_2} = \text{...}\text{mol/s}$$

Step 4
Calculate the total amount of N_2 that will diffuse out of the package in three months:

$$m_{N_2} = \text{...}\text{ g}$$

Verify that the pressure in the bag will not change appreciably due to N_2 diffusion.
Hint: Use the ideal gas law.

Chapter 12
Mass Transfer by Convection

Theory

For mass transfer by convection, the driving force is the concentration difference Δc expressed in mol/m^3 or mol fraction or pressure, and the resistance R is equal to $1/k_c A$, where k_c mass transfer coefficient and A surface area (m^2) perpendicular to the direction of transfer.

The mass transfer coefficient is calculated from relations of the form:

$$Sh = f(Re, Sc)$$

where

Sh = Sherwood number, $Sh = \dfrac{k_c L}{D_{ij}}$

Re = Reynolds number, $Re = \dfrac{Lv\rho}{\mu}$

Sc = Schmidt number, $Sc = \dfrac{\mu}{\rho D_{ij}}$

D_{ij} = diffusion coefficient of component i in component j, m^2/s
k_c = mass transfer coefficient, m/s or $mol/s\ m^2$ mol fraction or $mol/s\ m^2$ Pa
L = characteristic dimension, m
v = fluid velocity, m/s
μ = fluid viscosity, Pa s
ρ = fluid density, kg/m^3

To calculate the mass transfer coefficient:

1. Determine if the flow is natural or forced (free or forced convection).
2. Identify the geometry of the system.
3. Determine if the flow is laminar or turbulent (calculate the Reynolds number).

4. Select the appropriate relationship $Sh = f(Re, Sc)$.
5. Calculate Sh and solve for k_c.

Review Questions

Which of the following statements are true and which are false?

1. Mass transfer by convection is analogous to heat transfer by conduction.
2. The mass transfer coefficient may be expressed in different units which depend on the units that are used in the driving force.
3. The mass transfer coefficient depends on the physical properties of the fluid, the flow regime, and the geometry of the system.
4. The resistance to mass transfer by convection is proportional to the mass transfer coefficient.
5. A concentration boundary layer develops on a fluid flowing on a solid surface when the concentration of species "i" in the fluid is different from its concentration on the solid surface.
6. Concentration gradients exist in the concentration boundary layer.
7. The Schmidt number represents the ratio of mass diffusivity to momentum diffusivity.
8. The Lewis number represents the ratio of thermal diffusivity to mass diffusivity.
9. The Schmidt number relates the thickness of the hydrodynamic boundary layer to the thickness of the concentration boundary layer.
10. For dilute solutions the mass transfer coefficient for equimolar counterdiffusion is equal to the mass transfer coefficient for i through stagnant j.

Examples

Example 12.1

In a fruit packaging house, oranges are washed on a perforated belt conveyor by water sprays and dried in a stream of high speed air at room temperature before waxing. Calculate the mass transfer coefficient on the surface of an orange at 20°C if the air velocity is 10 m/s, the orange has a spherical shape with a diameter of 7 cm and the diffusivity of water vapor in air is $2.5 \times 10\text{-}5$ m2/s.

Solution

Step 1
Draw the process diagram:

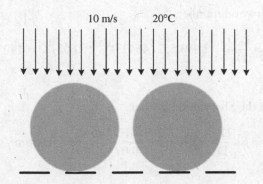

Step 2
State your assumptions:

- The moisture in the air is low so that the physical properties of air can be used:
- The oranges are apart from each other so that air flows around each one.

Step 3
Find the physical properties of air at 20°C:

$$\rho = 1.207 \text{g/m}^3$$

$$\mu = 1.82 \times 10^{-5} \text{kg/ms}$$

Step 4
Calculate the Reynolds number:

$$\text{Re} = \frac{D v \rho}{\mu} = \frac{(0.07 \,\text{m}) \, (10 \,\text{m/s}) \, \left(1.207 \,\text{kg/m}^3\right)}{1.82 \times 10^{-5} \,\text{kg/ms}} = 46423$$

Step 5
Calculate the Schmidt number:

$$\text{Sc} = \frac{\mu}{\rho D_{ij}} = \frac{1.82 \times 10^{-5} \,\text{kg/ms}}{\left(1.207 \,\text{kg/m}^3\right) \, \left(2.5 \times 10^{-5} \text{m}^2/\text{s}\right)} = 0.602$$

Step 6
Select a suitable equation of $\text{Sh} = f(\text{Re}, \text{Sc})$ for flow around a single sphere:

$$Sh = 2 + 0.552 \, Re^{0.53} \, Sc^{1/3}$$

Step 7
Calculate the Sherwood number:

$$Sh = \frac{k_c D}{D_{ij}} = 2 + 0.552 \, (46423)^{0.53} (0.602)^{1/3} = 140.6$$

Step 8
Calculate k_c from the Sherwood number:

$$k_c = Sh \frac{D_{ij}}{D} = 140.6 \frac{2.5 \times 10^{-5} \, m^2/s}{0.07 \, m} = 0.050 \, m/s$$

Exercises

Exercise 12.1

Instant coffee is dried in a spray dryer. Calculate the mass transfer coefficient on the surface of a coffee droplet that falls through the air in the spray dryer at the initial stages of drying if the diameter of the droplet is 0.5 mm, the relative velocity between the air and the droplet is 50 m/s, and the air temperature is 180°C. Assume that the droplet surface temperature is at 60°C.

Solution

Step 1
State your assumptions:

- The moisture in the air is low so that the physical properties of air can be used.
- The droplet has a spherical shape with constant diameter.

Step 2
Find the physical properties of air at the film temperature (the average temperature between droplet surface and bulk fluid):

$$\rho = \text{..} kg/m^3$$

$$\mu = \text{..} kg/ms$$

$$D_{ij} = \text{..} m^2/s$$

The diffusion coefficient of water vapor in air at 25°C and 1 atm is $2.6 \times 10^{-5} m^2/s$. Estimate the diffusion coefficient at 120°C using the relation $D_{ij} \propto T^{1.5}$ where T is degrees Kelvin.

Step 3
Calculate the Reynolds number:

$$Re = ...$$

Step 4
Calculate the Schmidt number:

$$Sc = ...$$

Step 5
Select a suitable equation of $Sh = f(Re, Sc)$ for flow around a single sphere:

$$Sh = ...$$

Step 6
Calculate the Sherwood number:

$$Sh = ...$$

Step 7
Calculate k_c from the Sherwood number:

$$k_c = ...$$

Exercise 12.2

Calculate the mass transfer coefficient for oxygen transfer from an air bubble rising in a non-agitated fermentation broth if the bubble diameter is 0.5 mm, the diffusion coefficient of O_2 in the broth is $2 \times 10^{-9} m^2/s$, the density and the viscosity of the broth are 1200 kg/m³ and 0.002 kg/m s, and the density of the gas is 1.1 kg/m³.

Solution

Step 1
State your assumptions: The bubble has a spherical shape with constant diameter.

Step 2
Select a suitable equation for mass transfer to small bubbles:

$$Sh = 2 + 0.31 \, Gr^{1/3} \, Sc^{1/3}$$

Step 3
Calculate the Grashof number:

$$Gr = \frac{D_b^3 \, \rho \, \Delta\rho \, g}{\mu^2} = \text{...}$$

Step 4
Calculate the Schmidt number:

$$Sc = \text{...}$$

Step 5
Calculate the Sherwood number:

$$Sh = \text{...}$$

Step 6
Calculate k_c from the Sherwood number:

$$k_c = \text{...}$$

Exercise 12.3

Several correlations are available in the literature for calculating mass transfer coefficients for different geometries. Three such correlations for flow past single spheres are:

$$Sh = 2 + 0.60 Re^{1/2} Sc^{1/2} \qquad\qquad Re < 70000$$
$$Sh = 2 + 0.552 Re^{053} Sc^{1/2} \qquad\qquad 1 < Re < 48000$$
$$Sh = 2 + \left(0.40 Re^{1/2} + 0.06 Re^{2/3}\right) Sc^{0.4} \qquad 3.5 < Re < 7600$$

Develop a spreadsheet program to evaluate and plot Sh vs. Re from these correlations for $1 < Re < 100$ and for $Sc = 0.6$ and $Sc = 2$. Repeat the calculation for $100 < Re$ 50000. Compare the results.

Exercise 12.4

Air flows over the surface of a solid tray 0.50 m x 0.50 m filled with a vegetable cut into small cubes that completely cover the tray. Calculate the rate of water vapor transfer from the surface of the tray to the air stream if the mass transfer coefficient is 0.03 kg/s m^2 Pa. The water vapor pressure is 12000 Pa at the surface of the vegetable and 5000 Pa at the bulk of the air stream.

Solution

Step 1
Draw the process diagram:

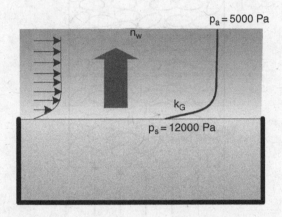

Step 2
State your assumptions:
The vegetable surface is covered with unbound moisture.

Step 3
Calculate the mass transfer rate:

$$n_w = k_G A(p_s - p_a) = \text{...}$$

Exercise 12.5

Air passes through a fixed bed of corn of 6 m^3 where the mass transfer coefficient is 0.05 kg/s m^2 Pa and the surface area per unit volume of the bed is 60 m^2/m^3. The water vapor pressure at the surface of the corn is 5000 Pa at the bottom of the bed and 5500 Pa at the top. Water vapor pressure for the bulk air is 1800 Pa at the inlet and 2500 Pa at the outlet of the bed. Calculate the rate of water vapor transfer from the corn to the air.

Solution

Step 1
Draw the process diagram:

Step 2
State your assumptions:The calculations will be based on steady state conditions.

Step 3
Select the equation to use:

$$n = k_G A \, \Delta P$$

i) Calculate the mean driving force Palong the bed. The logarithmic mean driving force should be used in this case:

$$\Delta P = \frac{(p_{s1} - p_{a1}) - (p_{s2} - p_{a2})}{\ln\dfrac{p_{s1} - p_{a1}}{p_{s2} - p_{a2}}} = \dots\dots\dots\dots\dots\dots\dots\dots\dots\dots\dots\dots$$

ii) Calculate the total external surface area of the solid particles:

$$A = \dots\dots\dots\dots\dots\dots\dots\dots\dots\dots\dots\dots$$

iii) Calculate the mass transfer rate:

$$n = \dots\dots\dots\dots\dots\dots\dots\dots\dots\dots\dots\dots$$

Chapter 13
Unsteady State Mass Transfer

Theory

The unidirectional unsteady state mass transfer equation in the x-direction is given by Fick's 2nd law:

$$\frac{\partial C}{\partial t} = D_{ij} \frac{\partial^2 C}{\partial x^2}$$

The solutions of the unsteady state mass transfer equation are the same with the solutions of the unsteady state heat transfer equation given in Tables 10.1, 10.2, and 10.4, with the following correspondence between heat and mass transfer variables:

where

Bi_m = Biot number for mass transfer
C_{Fi} = species concentration in the fluid at the interface, mol/m^3, kg/m^3

Table 13.1 Relations between heat and mass transfer parameters

Heat transfer	Mass transfer
$\dfrac{T_e - T}{T_e - T_o}$	$\dfrac{C_s - C}{C_s - C_o}$
$Fo = \dfrac{\alpha t}{L^2}$	$Fo_m = \dfrac{D_{ij} t}{L^2}$
$Bi = \dfrac{hL}{k}$	$Bi_m = \dfrac{m k_c L}{D_{ij}}$
$\dfrac{x}{2\sqrt{\alpha t}}$	$\dfrac{x}{2\sqrt{D_{ij} t}}$
$\dfrac{h\sqrt{\alpha t}}{k}$	$\dfrac{m k_c \sqrt{D_{ij} t}}{D_{ij}}$
	$m = \dfrac{C_{Fi}}{C_s}$ if $Bi_m \gg 1$, $C_{Fi} = C_{FL}$

C_{FL} = species concentration in the bulk of the fluid, mol/m^3, kg/m^3
C_s = species concentration in the medium at the surface of the solid, mol/m^3, kg/m^3
C_o = initial concentration of the species in the solid, mol/m^3, kg/m^3
C = concentration of the species in the solid at time t and point x, mol/m^3, kg/m^3
D_{ij} = mass diffusivity, m^2/s
Fo_m = Fourier number for mass transfer
k_c = mass transfer coefficient, m/s
L = characteristic dimension (half thickness of the plate, radius of a cylinder or sphere), m
m = equilibrium distribution coefficient
x = distance of the point from the center plane of the plate, or from the surface in the case of semi-infinite body, m
t = time, s

To calculate the concentration as a function of time and position in a solid we follow the same procedure as for calculating the temperature as a function of time and position in a solid:

1. Identify the geometry of the system. Determine if the solid can be considered as plate, infinite cylinder, or sphere.
2. Determine if the surface concentration is constant. If not, calculate the Biot number and decide the relative importance of internal and external resistance to mass transfer.
3. Select the appropriate equation given for unsteady state heat transfer (Tables 10.1, 10.2, 10.4).
4. Calculate the Fourier number.
5. Find the concentration by applying the selected equation (if Fo > 0.2, calculate the temperature either using only the first term of the series solution or using Heisler or Gurnie-Lurie charts).

Review Questions

Which of the following statements are true and which are false?

1. If the concentration at any given point of a body changes with time, unsteady state mass transfer occurs.
2. Fick's 2nd law is used in unsteady state mass transfer problems.
3. Gurnie-Lurie and Heisler charts are valid for unsteady state mass transfer problems.
4. The Heisler chart can be used to find the concentration at any point in a body if the Fourier and Biot numbers are known.
5. Gurnie-Lurie charts are used when Fo < 0.2.

6. The distribution coefficient of the diffusing component between fluid and solid is used in unsteady state mass transfer problems.
7. The required time for a certain change in concentration due to diffusion is proportional to the square of the thickness of the body.
8. If only one surface of an infinite slab is exposed to the diffusing substance, half the thickness of the slab is used as characteristic dimension.
9. Diameter is used as the characteristic dimension of a sphere for the solution of unsteady state mass transfer problems.
10. The solutions of an infinite cylinder and an infinite slab are combined to find the concentration distribution in a cylinder of finite length.

Examples

Example 13.1

A food item $2 \times 20 \times 20$ cm with 200 kg water/m^3 moisture content is exposed to a high speed hot air stream in a dryer. The surface of the food immediately attains moisture equal to 50 kg water/m^3. This surface condition is maintained thereafter. If the diffusivity of moisture in the food is $1.3 \times 10^{-9} m^2/s$, calculate: a) the moisture content at the center after 5 hours, b) how much time is required to bring the moisture content at the center to a value of 100 kg water/m^3, and c) the mass average moisture content at this time.

Solution

Step 1
Draw the process diagram:

Step 2
State your assumptions:

- Moisture is initaially uniform throughout the body.
- Moisture transfer in the solid is diffusion controlled.
- The volume of the solid is constant.
- The diffusivity is constant.

Step 3
Identify the geometry of the system.
The shape of the food can be considered as an infinite slab since the thickness is much smaller than the other two dimensions.

Step 4

i) Select the appropriate equation to use.

Since the surface concentration is constant, the equation is (see Tables 10.1 and 13.1).

$$\frac{C_s - C}{C_s - C_o} = \frac{4}{\pi}\sum_{n=0}^{\infty}\frac{(-1)^n}{2n+1}\cos\left(\frac{(2n+1)\pi x}{2L}\right)\exp\left(-\frac{(2n+1)^2\pi^2}{4}Fo_x\right) \quad (13.1)$$

ii) Calculate the Fourier number

$$Fo_x = \frac{Dt}{L_x^2} = \frac{(1.3\times10^{-9}\,\mathrm{m^2/s})\,(5\times3600\,\mathrm{s})}{(0.01\ \mathrm{m})^2} = 0.234$$

Since $Fo_x > 0.2$ the first term of the series in eqn (13.1) is enough. Alternatively, the Heisler chart can be used.

Step 5

i) Find the concentration at the center using eqn (3.1) above:

$$\frac{C_s - C}{C_s - C_o} = \frac{4}{\pi}\sum_{n=0}^{\infty}\frac{(-1)^n}{2n+1}\cos\left(\frac{(2n+1)\pi x}{2L}\right)\exp\left(-\frac{(2n+1)^2\pi^2}{4}Fo_x\right)$$

$$= \frac{4}{\pi}\frac{(-1)^0}{(2\times0+1)}\cos\frac{(2\times0+1)\times\pi\times0}{2\times0.01}\exp\left(-\frac{(2\times0+1)^2\pi^2}{4}\times0.234\right)$$

$$= 0.715$$

and

$$C = C_s - 0.715(C_s - C_o) = 50 - 0.715\,(50 - 200) = 157 \text{ kg water/m}^3$$

ii) Find the concentration at the center using the Heisler chart for an infinite slab (Fig A.5):

• Find the value of $Fo = 0.234$ on the x-axis.
• Find the curve with $1/Bi = 0$.
• Read the dimensionless concentration on the y-axis as:

$$\frac{C_s - C}{C_s - C_o} = 0.7$$

$$C = C_s - 0.7(C_s - C_o)$$

$$= 50 - 0.7\,(50 - 200) = 155 \text{ kg water/m}^3$$

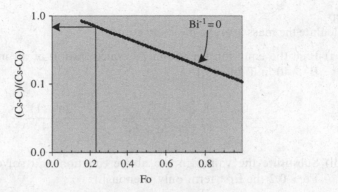

Step 6

Find the required time for the moisture content to drop to 100 kg water/m^3 at the center.

The time will be found from the Heisler chart:

Calculate the dimensionless concentration

$$\frac{C_s - C}{C_s - C_o} = \frac{50 - 100}{50 - 200} = 0.33$$

On the Heisler chart:

- Find on the y-axis the value
- $(C_s - C)/(C_s - C_0) = 0.33$.
- Find the curve $1/Bi = 0$.
- Read the Fourier number on the x-axis as $F_o = 0.54$.
- Find the time from the Fourier number as:

$$t = Fo\frac{L_x^2}{D_{ij}} = 0.54\frac{0.01^2\,m^2}{1.3 \times 10^{-9}\,m^2/s} = 41538\ s$$

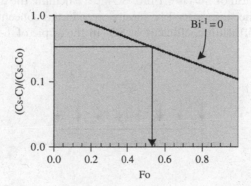

Step 7
Calculate the mass average moisture content.

   i) Find the equation for the mean concentration of an infinite slab with Bi > 40 in Table 10.1:

$$\frac{C_s - C_m}{C_s - C_o} = \frac{8}{\pi^2} \sum_{n=0}^{\infty} \frac{1}{(2n+1)^2} \exp\left(-\frac{(2n+1)^2 \pi^2}{4} Fo\right)$$

   ii) Substitute the values in the above equation and solve for C_m (since Fo > 0.2 the first term only is enough).

$$\frac{C_s - C_m}{C_s - C_o} = \frac{8}{\pi^2} \sum_{n=0}^{\infty} \frac{1}{(2n+1)^2} \exp\left(-\frac{(2n+1)^2 \pi^2}{4} Fo\right) =$$

$$= \frac{8}{\pi^2} \frac{1}{(2 \times 0 + 1)^2} \exp\left(-\frac{(2 \times 0 + 1)^2 \pi^2}{4} \times 0.54\right) = 0.214$$

and

$$C_m = C_s - 0.214(C_s - C_o) = 50 - 0.214\,(50 - 200) = 82.1 \ \text{kg water/m}^3$$

Exercises

Exercise 13.1

Apple slices 4 mm thick are exposed to sulphur fumes in a sulphuring house before drying. If the Biot number for mass transfer is higher than 40 and the surface concentration of SO_2 is 0.1 mol SO_2/m^3, calculate the mass average SO_2 concentration in the slices after 1 h. Assume zero initial concentration of SO_2 in the apple and a diffusion coefficient of SO_2 in the apple of $1 \times 10^{-9} m^2/s$.

Solution

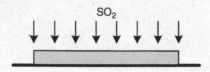

Step 1
State your assumptions:

- Mass transfer through the sides of the slice is negligible (the solution for an infinite slab can be applied).
- Only the top of the slice is exposed to sulphur fumes.

Step 2
Calculate the Fourier number.
Since the food is exposed to the fumes *from one side* the characteristic dimension is equal to the thickness of the slice. Therefore:

$$Fo = \text{...}$$

Step 3
Find the equation for the mean concentration of a plate with $Bi > 40$ in Table 10.1. Substitute the values in the equation and solve for C_m:

$$\frac{C_s - C_m}{C_s - C_o} = \text{...}$$

and

$$C_m = \text{...}$$

Exercise 13.2

Calculate how long it will take for the concentration on the non-exposed surface of the apple slice of Exercise 13.1 to increase by 1% of the initial concentration difference.

Solution

Step 1
Identify the geometry of the system.
As long as the concentration change on the non-exposed surface of the apple slice is less than 1% of the initial concentration difference, the slice can be treated as a semi-infinite body.

Step 2
Select the appropriate equation from Table 10.4.
Since the surface concentration is constant, the equation for a semi-infinite body with uniform initial concentration and constant surface concentration will be used:

$$\frac{C_s - C}{C_s - C_o} = \text{erf}\left(\frac{x}{2\sqrt{Dt}}\right)$$

Step 3
The mass penetration depth is given by (as in heat transfer):

$$\frac{x}{2\sqrt{Dt}} = \text{...............................}$$

Step 4
Solve for t, substitute values, and calculate t:

$$t = \text{...}$$

Comment: This result shows that for the first .... the body can be
treated as a semi-infinite body.

Exercise 13.3

A piece of feta cheese 15cm × 15cm × 10cm is immersed in NaCl brine. Calculate the NaCl concentration in the center of the piece and the mass average
concentration after 10 days if the concentration at the surface immediately
attains a value of $130 kgNaCl/m^3$ and remains constant during the brining
process, the cheese did not contain any salt at the beginning, and the diffusion
coefficient of NaCl in the cheese is $3 \times 10^{-10} m^2/s$.

Solution

Step 1
Draw the process diagram:

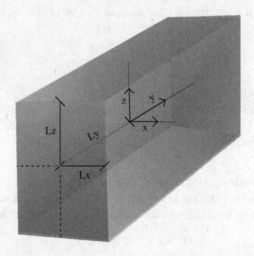

Step 2
State your assumptions:

- NaCl transfer in the cheese is diffusion controlled.
- The volume of the cheese remains constant.

Step 3
Define the shape of the object.
The cheese is a rectangular slab. The concentration in the center is affected by the mass transferred from the three directions x, y, and z. The contribution from each direction has to be calculated separately and the combined effect will be calculated at the end.

Step 4
Select the appropriate equation to use.
Since the concentration at the surface is constant, external resistance is considered negligible. Therefore, the solution for the concentration at the center will be:

$$\frac{C_s - C}{C_s - C_o} = \frac{4}{\pi} \sum_{n=0}^{\infty} \frac{(-1)^n}{2n+1} \cos\left(\frac{(2n+1)\pi x}{2L}\right) \exp\left(-\frac{(2n+1)^2 \pi^2}{4} Fo\right)$$

Step 5
Calculate mass transfer in x, y, and z directions.

1) Mass transferred in the x-direction:

   i) Calculate the Fourier number Fo_x for the x-direction:

   $$Fo_x = \dots\dots\dots\dots\dots\dots\dots\dots\dots\dots\dots\dots\dots\dots\dots\dots\dots$$

Since Fo_x the Heisler chart cannot be used. More than one term of the sum in the above equation has to be used. Three terms will be enough for this problem.

   ii) Calculate the dimensionless concentration:

$$\frac{C_s - C_x}{C_s - C_o} = \frac{4}{\pi} \sum_{n=0}^{\infty} \frac{(-1)^n}{2n+1} \cos\left(\frac{(2n+1)\pi x}{2L_x}\right) \exp\left(-\frac{(2n+1)^2 \pi^2}{4} Fo_x\right) \approx$$

$$= \frac{4}{\pi} \left[\begin{array}{l} \dfrac{(-1)^0}{2\times 0+1}\cos\left(\dfrac{(2\times 0+1)\pi\times 0}{2\times 0.075}\right)\exp\left(-\dfrac{(2\times 0+1)^2\pi^2}{4}\dots\right) \\[2mm] +\dots\dots\dots\dots\dots\dots\dots\dots\dots \\[4mm] \dots\dots\dots\dots\dots\dots\dots\dots\dots\dots\dots\dots \\[4mm] +\dots\dots\dots\dots\dots\dots\dots\dots\dots\dots\dots\dots\dots\dots \end{array} \right] =$$

$$= \dots\dots\dots\dots\dots\dots\dots$$

2) Mass transferred in the y-direction:

   i) Calculate the Fourier number Fo_y for the y-direction:

$$Fo_y = \dots\dots\dots\dots\dots\dots\dots\dots\dots\dots\dots\dots\dots\dots\dots$$

   ii) Calculate the dimensionless concentration:

$$\frac{C_s - C_y}{C_s - C_o} = \frac{4}{\pi} \sum_{n=0}^{\infty} \frac{(-1)^n}{2n+1} \cos\left(\frac{(2n+1)\pi y}{2L_y}\right) \exp\left(-\frac{(2n+1)^2 \pi^2}{4} Fo_y\right) \approx$$

$$= \frac{4}{\pi} \left[\begin{array}{l} = \dots\dots\dots\dots\dots\dots\dots\dots\dots\dots\dots\dots\dots\dots\dots\dots \\[2mm] + \dots\dots\dots\dots\dots\dots\dots\dots\dots\dots\dots \\[4mm] \dots\dots\dots\dots\dots\dots\dots\dots\dots\dots\dots\dots \\[4mm] + \dots\dots\dots\dots\dots\dots\dots\dots\dots\dots\dots\dots\dots \end{array} \right] =$$

$$= \dots\dots\dots\dots\dots\dots\dots$$

3) Mass transferred in the z-direction:

   i) Calculate the Fourier number Fo_z for the z-direction:

$$Fo_z = \dots\dots\dots\dots\dots\dots\dots\dots\dots\dots\dots\dots\dots\dots\dots$$

   ii) Calculate the dimensionless concentration.

$$\frac{C_s - C_z}{C_s - C_o} = \frac{4}{\pi} \sum_{n=0}^{\infty} \frac{(-1)^n}{2n+1} \cos\left(\frac{(2n+1)\pi z}{2L_z}\right) \exp\left(-\frac{(2n+1)^2 \pi^2}{4} Fo_z\right) \approx$$

$$= \frac{4}{\pi} \left[\begin{array}{l} \dots\dots\dots\dots\dots\dots\dots\dots\dots\dots\dots\dots\dots\dots\dots\dots \\[2mm] + \dots\dots\dots\dots\dots\dots\dots\dots\dots\dots\dots \\[4mm] \dots\dots\dots\dots\dots\dots\dots\dots\dots\dots\dots\dots \\[4mm] + \dots\dots\dots\dots\dots\dots\dots\dots\dots\dots\dots\dots\dots \end{array} \right] =$$

$$= \dots\dots\dots\dots\dots\dots\dots$$

4) The combined effect of mass transferred in the x, y, and z directions is:

$$\frac{C_s - C_{xyz}}{C_s - C_o} = \left(\frac{C_s - C_x}{C_s - C_o}\right)\left(\frac{C_s - C_y}{C_s - C_o}\right)\left(\frac{C_s - C_z}{C_s - C_o}\right)$$

$$= \dots\dots\dots\dots\dots\dots\dots\dots\dots$$

Calculate the concentration at the center as:

$$\frac{\dots\dots\dots - C_{xyz}}{\dots} = \dots\dots\dots \rightarrow C_{xyz} =$$

Step 6
The solution for the mass average concentration is:

$$\frac{C_s - C_m}{C_s - C_o} = \frac{8}{\pi^2}\sum_{n=0}^{\infty}\frac{1}{(2n+1)^2}\exp\left(-\frac{(2n+1)^2\pi^2}{4}Fo\right)$$

1) Mass transferred in the x-direction:

$$\frac{C_s - C_{mx}}{C_s - C_o} = \frac{8}{\pi}\sum_{n=0}^{\infty}\frac{1}{(2n+1)^2}\exp\left(-\frac{(2n+1)^2\pi^2}{4}Fo_x\right) =$$

$$= \frac{8}{\pi}\left[\begin{array}{l}\dots\dots\dots\dots\dots\dots\dots\dots\dots\dots\dots\dots\dots\dots\dots\dots\dots \\ +\dots\dots\dots\dots\dots\dots\dots\dots \\ \\ \dots\dots\dots\dots\dots\dots\dots\dots\dots\dots \\ \\ +\dots\dots\dots\dots\dots\dots\dots\dots\dots\dots\dots\dots\dots \end{array}\right] =$$

$$= \dots\dots\dots\dots\dots\dots\dots$$

2) Mass transferred in the y-direction:

$$\frac{C_s - C_{my}}{C_s - C_o} = \frac{8}{\pi} \sum_{n=0}^{\infty} \frac{1}{(2n+1)^2} \exp\left(-\frac{(2n+1)^2 \pi^2}{4} Fo_y\right) \approx$$

$$= \frac{8}{\pi} \left[\begin{array}{c} \dots\dots\dots\dots\dots\dots\dots\dots\dots\dots + \dots\dots\dots\dots\dots\dots\dots\dots\dots \\ \\ \dots\dots\dots\dots\dots\dots\dots\dots\dots + \dots\dots\dots\dots\dots\dots\dots\dots \end{array} \right] =$$

$$= \dots\dots\dots\dots\dots\dots\dots\dots\dots$$

3) Mass transferred in the z-direction:

$$\frac{C_s - C_{mz}}{C_s - C_o} = \frac{8}{\pi} \sum_{n=0}^{\infty} \frac{1}{(2n+1)^2} \exp\left(-\frac{(2n+1)^2 \pi^2}{4} Fo_z\right) \approx$$

$$= \frac{8}{\pi} \left[\begin{array}{c} \dots\dots\dots\dots\dots\dots\dots\dots\dots\dots + \dots\dots\dots\dots\dots\dots\dots\dots\dots \\ \\ \dots\dots\dots\dots\dots\dots\dots\dots\dots + \dots\dots\dots\dots\dots\dots\dots\dots \end{array} \right] =$$

$$= \dots\dots\dots\dots\dots\dots\dots\dots\dots$$

4) The combined effect of mass transferred in the x, y, and z directions is:

$$\frac{C_s - C_{mxyz}}{C_s - C_o} = \left(\frac{C_s - C_{mx}}{C_s - C_o}\right)\left(\frac{C_s - C_{my}}{C_s - C_o}\right)\left(\frac{C_s - C_{mz}}{C_s - C_o}\right)$$

$$= \dots\dots\dots\dots\dots\dots\dots\dots$$

Calculate the mass average concentration as:

$$\frac{\dots\dots\dots\dots - C_{mxyz}}{\text{———————————————}} = \dots\dots\dots\dots\dots \rightarrow$$

$$C_{mxyz} = \dots\dots\dots\dots\dots\dots\dots$$

Exercise 13.4

To remove the excess salt from salt-stock cucumbers in a pickle factory, the cucumbers are immersed in several changes of fresh water. Calculate the average NaCl concentration in a cucumber after 5 hours of immersion in fresh water if the cucumber has 2 cm diameter and 6 cm length, the initial NaCl content is 100 kg NaCl/m^3, and the diffusivity of NaCl in the cucumber is 1×10^{-9} m^2/s. Assume that the change of water is continuous so that the salt concentration in the water is zero.

Solution

Step 1
Draw the process diagram:

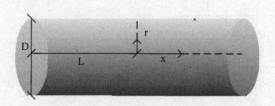

Step 2
State your assumptions:

- The external resistance to mass transfer is negligible so that ....
  ....
- Diffusion through the ends of the pickle is negligible
- ....

Step 3
Identify the geometry of the system.
The pickle will be treated as an infinitely long cylinder:

Step 4
Select the equation to use.
The solution for the mean concentration for a cylinder with constant surface concentration will be applied for the cylindrical surface, that is:

$$\frac{C_s - C_m}{C_s - C_o} = \frac{4}{\pi^2} \sum_{n=1}^{\infty} \frac{\pi^2}{R^2 \delta_n^2} \exp\left(-R^2 \delta_n^2 Fo_r\right) =$$

$$= \frac{4}{\pi^2} \left(1.7066 \exp(-5.783 Fo_r) + 0.324 \exp(-30.5 Fo_r) \right.$$

$$\left. + 0.132 \exp(-74.9 Fo_r) + ... \right)$$

Step 5
Calculate the Fourier number:

$$Fo_r = ...$$

Since $Fo_r > 0.2$, more than one term of the sum in the above equation will be used. Two terms are enough for this case. Verify that even the contribution of the 2nd term is small.

Step 6

Substitute the values in the above equation and solve for C_m:

$$\frac{C_s - C_m}{C_s - C_o} \approx \frac{4}{\pi^2}\left(1.7066\exp(-5.783Fo_r) + 0.324\exp(-30.5Fo_r)\right) =$$

$$= \text{...} =$$

$$= \text{........................}$$

and

$$C_m = \text{..}$$

Exercise 13.5

The moisture content of the soil in a field is 200 kg water/m³, when suddenly dry air starts blowing. Calculate the moisture content of the soil 5 cm below the surface after 48 hours if the diffusion coefficient of water in the soil is $1.3 \times 10^{-8}\,m^2/s$ and the equilibrium moisture content of the soil in contact with the air is 100 kg water/m³. Assume that the surface immediately equilibrates with the air.

Solution

Step 1

Draw the process diagram:

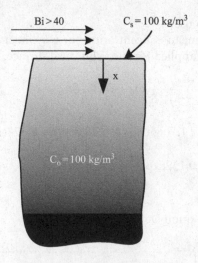

Step 2
State your assumptions:

- Initially, the moisture content in the soil is uniform.
- The moisture transfer in the soil is diffusion controlled.

Step 3
Identify the shape of the object.Ground can be treated as a semi-infinite body.

Step 4
Select the equation to use.The equation for unsteady state in a semi-infinite body with constant surface concentration will be used:

$$\frac{C_s - C}{C_s - C_o} = \mathrm{erf}\left(\frac{x}{2\sqrt{Dt}}\right)$$

i) Calculate:

$$\frac{x}{2\sqrt{Dt}} = \text{\underline{\hspace{4cm}}} = \ldots\ldots\ldots$$

ii) Find the value of $\mathrm{erf}\left(\dfrac{x}{2\sqrt{\alpha t}}\right)$ from Table A.5

iii) Substitute values in the original equation and calculate:

$$\frac{C_s - C}{C_s - C_o} = \ldots\ldots\ldots\ldots\ldots\ldots\ldots\ldots\ldots\ldots$$

$$\rightarrow C = \ldots\ldots\ldots\ldots\ldots\ldots\ldots\ldots\ldots$$

Exercise 13.6

Solve Example 13.1 using the spreadsheet program *Mass Transfer-Negligible Surface Resistance.xls.*

Solution

Step 1
On the spreadsheet *Mass Transfer-Negligible Surface Resistance.xls,* go to the sheet "slab." Turn the "SWITCH" OFF (set the value in cell G1 equal to 0 and press ENTER).

Step 2
Insert the parameter values in the yellow cells using: half thickness in the x-direction $L_x = 0.01$m (since the slice is exposed to the air from both sides,

L_x is taken equal to half thickness), half thickness in the y-direction $L_y = 0.10$m, half thickness in the z-direction $L_z = 0.10$m, $x = 0$, $y = 0$ and $z = 0$ (since the concentration will be calculated at the center), initial moisture content $C_o = 200$kgwater/m^3, surface concentration $C_s = 50$kgwater/m^3. Also insert the given value for the diffusivity.

Step 3
Set the value of the time step to 120 s (or some other value) in cell F4.

Step 4
Turn the "SWITCH" ON (set the value in cell G1 equal to 1 and press ENTER).

Step 5
Iterate by pressing F9 until the moisture in cell K20 reaches the value 100.

Step 6
Read the mass average moisture content in cell K40. Observe if the average moisture is affected by the mass transferred from the y and z directions.

Step 7
To see how the mass average concentration would have decreased if the slice was exposed to the air from one side only:

i) Turn the "SWITCH" OFF (set the value in cell G1 equal to 0 and press ENTER).
ii) Set the value of half thickness in the x-direction to $L_x = 0.02$m.
iii) Set the value of the coordinates of the point to $x = 0, y = 0$ and $z = 0$.
vi) Turn the "SWITCH" ON (set the value in cell G1 equal to 1 and press ENTER).
v) Iterate by pressing F9 until the mass average moisture in cell K40 reaches the same value as in 6 above.
vi) Read the time in cell F3. Compare the two time values.

Exercise 13.7

Small identical cubes 2cm × 2cm × 2cm of a cheese-like product were dry-salted (covered with dry NaCl crystals). From time to time, one cube was taken out and analyzed for NaCl and moisture content. Calculate the diffusion coefficient of NaCl in the cheese if the following average NaCl concentration vs. time data were obtained.

Time, h	kg NaCl/m^3
0	3
2	122
6	163
10	194
24	206
30	215
48	221
60	221

Solution

Step 1
State your assumptions:

....

Step 2
On the spreadsheet *Mass Transfer-Negligible Surface Resistance.xls,* go to the sheet "slab." Turn the "SWITCH" OFF (set the value in cell G1 equal to 0 and press ENTER).

Step 3
Insert the parameter values in the yellow cells using: half thickness in the x-direction $L_x = 0.01$ m, half thickness in the y-direction $L_x = 0.01$m, half thickness in the z-direction $L_z = 0.01$m, $x = 0$, $y = 0$ and $z = 0$, initial NaCl content $C_o = 3 kgNaCl/m^3$, surface concentration $C_s = 221 kgNaCl/m^3$.

Step 4
Insert a first trial value for the diffusivity.

Step 5
Set the value of the time step to 1000 s (or some other value) in cell F4.

Step 6
Turn the "SWITCH" ON (set the value in cell G1 equal to 1 and press ENTER).

Step 7
Iterate by pressing F9 until the mass average NaCl concentration in cell K40 reaches the value of 221 (final experimental concentration).

Step 8
Calculate the mean % deviation between the experimental values and the theoretically predicted values of a mass average concentration of NaCl from:

$$\text{Deviation } \% = \frac{100}{N} \sum_{1}^{N} \frac{|C_{exp} - C_{theor}|}{C_{exp}}$$

Step 9
Turn the "SWITCH" OFF (set the value in cell G1 equal to 0 and press ENTER).

Step 10
Insert a second trial value for the diffusivity in cell B38.

Step 11
Turn the "SWITCH" ON (set the value in cell G1 equal to 1 and press ENTER)

Step 12
Iterate by pressing F9 until the mass average NaCl concentration in cell K40 reaches the value 221.

Step 13
Calculate the mean % deviation between the experimental values and the theoretically predicted values of a mass average concentration of NaCl as above.

Step 14
Repeat steps 9 to 13 until a minimum deviation % is reached.

Exercise 13.8

Solve Exercises 13.1, 13.2, 13.3, and 13.4 using the spreadsheet *Mass Transfer-Negligible Surface Resistance.xls*.

Chapter 14
Pasteurization and Sterilization

Review Questions

Which of the following statements are true and which are false?

1. The decimal reduction time D is the heating time in min at a certain temperature required for the number of viable microbes to be reduced to 10% of the original number.
2. The z value is the temperature increase required for a ten-fold decrease in D.
3. Thermal death time is the heating time required to give commercial sterility.
4. Thermal death time does not depend on the initial microbial load.
5. The D value does not depend on the initial microbial load.
6. The D value of a microorganism is independent of the food item.
7. The D value of a microbe is a measure of the thermal resistance of the microbe.
8. If the number of microbes in a process has to be reduced from an initial load of 10^6 to a final 10^{-4}, the required thermal death time will be 10D.
9. If the number of microbes in a canned product is reduced from 10^3 to 10^{-4}, it means that 1 can in 100000 may be spoiled.
10. As the process temperature increases, the thermal death time increases.
11. A 10D process is usually applied as a minimum heat treatment for Clostridium botulinum.
12. Typical z values are 5.5 °C for vegetative cells, 10 °C for spores, and 22 °C for nutrients.
13. A $D_{121.1} = 0.21$ min is usually assumed for Clostridium botulinum.
14. The accepted risk for Clostridium botulinum is 10^{-12}.
15. The accepted spoilage probability for mesophilic spoilage microorganisms is usually 10^{-5}.
16. The accepted spoilage propability for thermophilic spoilage microorganisms upon incubation after processing is usually 10^{-2}.
17. The slowest heating point of a can filled with a liquid food is the geometric center of the can.

18. The worst case scenario for calculating the lethal effect in a holding tube for a Newtonian liquid is to assume that the residence time is half the mean residence time.
19. In calculating the lethal effect in a solid particulate flowing in a two-phase flow in a holding tube, the convective heat transfer resistance at the surface of the particulate can be neglected.
20. The Ball Formula method is an alternative method to the general method for calculating the lethal effect of a sterilization process.
21. For the purpose of thermal processing, foods are divided into low acid foods (pH > 4.5) and high acid foods (pH < 4.5).
22. High acid foods need more heat treatment than low acid foods.
23. Clostridium botulinum is of concern in high acid foods.
24. Spoilage microorganisms is the major concern in low acid foods.
25. Thermal treatment of high acid foods is usually carried out at temperatures $\leq 100\,^{\circ}C$.

Examples

Example 14.1

The decimal reduction time D at $121\,^{\circ}C$ (D_{121}) and the value z for a thermophilic spore in whole milk were determined experimentally to be equal to 30 s and $10.5\,^{\circ}C$ respectively. Calculate the D value at $150\,^{\circ}C$ (D_{150}).

Solution

Step 1
The effect of temperature on D is given by:

$$D_{T_1} = D_{T_2} 10^{\frac{T_2 - T_1}{z}} \tag{14.1}$$

Step 2
Substitute values in eqn (14.1) and calculate D_{150}:

$$D_{150} = 0.5 \times 10^{\frac{121 - 150}{10.5}} = 0.000865\,\text{min}$$

Example 14.2

Determine the required heating time at $121\,^{\circ}C$, F_{121} value, in the case of Example 14.1 for a 9 log cycles population reduction.

Solution

Step 1

The required F value at temperature T for a certain population reduction is given by:

$$F_T = D_T \log \frac{N_o}{N} \tag{14.2}$$

e.g., for 12 log cycles population reduction, $F = 12D$.
Equation (14.2) can be used to convert log cycles of reduction into F_T (min at T) if D_T is known.

Step 2
Substitute values in eqn (14.2) and find F_{121}:

$$F_{121} = D_{121} \log \frac{N_o}{N} = 0.5 \times 9 = 4.5 \text{ min}$$

Thus the sterilizing value or the lethality for this process is $F = 4.5$ min.

Example 14.3

Find two equivalent processes, at 100 °C and 150 °C, which will deliver the same lethality as the required F_{121} value of 4.5 min calculated in the previous example.

Solution

Step 1
State your assumptions: All thermal treatment is delivered at a constant temperature of 100 °C or 150 °C.

Step 2
The delivered F value by a process at temperature T equivalent to a reference process F_R at T_R is given by:

$$F_T = F_R \times 10^{\frac{T_R - T}{z}} \tag{14.3}$$

Step 3
Substitute values in eqn (14.3) and find F_T:

$$F_{100} = F_{121} \times 10^{\frac{121 - T}{z}} = 4.5 \times 10^{\frac{121 - 100}{10.5}} = 450 \text{ min}$$

$$F_{150} = F_{121} \times 10^{\frac{121 - T}{z}} = 4.5 \times 10^{\frac{121 - 150}{10.5}} = 0.0078 \text{ min}$$

Example 14.4

The activation energy for vitamin C thermal destruction for 11.2 °C. Brix grapefruit juice was calculated to be equal to 4.98 kcal/mol (Ref. 12), based on k values between 61 °C and 96 °C. Calculate the z value for vitamin C thermal destruction in grapefruit juice.

Solution

The relationship between E_a and z is:

$$z = \frac{2.303\,RT^2}{E_a}$$

or as suggested in (Ref. 13):

$$z = \frac{2.303\,RT_{min}T_{max}}{E_a}$$

Substitute values and calculate z:

$$z = \frac{2.303\,RT_{min}T_{max}}{E_a} = \frac{2.303(1.9872\,cal/mol\,K)\,(273+61)K\,(273+96)K}{4980\,cal/mol}$$

$$= 113.3\,K\,or\,°C$$

Example 14.5

Calculate the F_o value for sterilization of cream in a can with the following time-temperature history

Time (min)	Temperature (°C)
0	50
3	80
5	100
8	115
11	119
13	121
14	119
15	110
17	85
19	60
20	50

Solution

Step 1
Since the F_o value is sought, 121.1 °C and 10 °C will be used as reference temperature and z value respectively. The equivalent time at 121.1 °C of the process will be calculated using the General method with numerical integration:

$$F_o = \int_0^t 10^{\frac{T-121.1}{10}} dt \approx \sum_0^t 10^{\frac{T-121.1}{10}} \Delta t \qquad (14.4)$$

Step 2
Use the spreadsheet program *General method.xls*. Insert the data and read the result of the numerical integration of eqn (14.4). Observe the contribution of the heating phase and the cooling phase on F_o.

Exercises

Exercise 14.1

The reaction velocity constant k for vitamin C destruction at 121 °C was found equal to 0.00143 min^{-1}. Express it in terms of decimal reduction time.

Solution

Step 1
The relationship between k and D is:

$$D_{121} = \frac{2.303}{k}$$

Step 2
Substitute k and calculate D:

$$D_{121} = \text{.....................}$$

Exercise 14.2

The reaction velocity constant k for vitamin C thermal destruction for 11.2 °C.Brix grapefruit juice at 95 °C is 0.002503 min^{-1} (Ref. 12). Calculate the time at this temperature for 10% destruction of vitamin C.

Solution

Step 1
Calculate D from k:

$$D_{95} = \text{.....................}$$

Step 2
Calculate the required time from an equation analogous to eqn (14.2) of
Example 14.2 above.

$$t = D_{95}\log\frac{C_o}{C} = \dots$$

Exercise 14.3

The following time-temperature history was recorded during pasteurization
of grapefruit juice. Calculate the thermal destruction of vitamin C due to
pasteurization.

Time (s)	Temperature (°C)
0	20
5	50
8	65
11	75
13	82
18	90
30	90
32	80
35	60
40	45
50	30

Solution

Step 1
Find the equivalent time of the process at 95 °C using the General method with
numerical integration as in step 1 of Example 14.5. Use the z value found in
Example 14.4. Use the spreadsheet program *General method.xls*. Insert the data
and read the result.

Step 2
Use the D_{95} value found in Exercise 14.2 and calculate C/C_o from a relationship
analogous to eqn (14.2) of Example 14.2.

$$\frac{C}{C_o} = \dots$$

Exercise 14.4

Experimental data from various sources presented in Ref. 12 indicate that the
energy of activation for thermal destruction of vitamin B_1 in certain vegetables

and meat is in the range of 26–28 kcal/mol based on k values between 70.5 °C and 149 °C with $k_{109°C} = 0.0063$ min^{-1}. Calculate the loss of vitamin B$_1$ due to thermal destruction in the case of cooking the meat for 90 min at 100 °C or alternatively for 30 min in a pressure cooker at 115 °C

Solution

Step 1
Calculate the z value and D$_{109}$ value:

$$z = \dots\dots\dots\dots\dots\dots\dots\dots\dots\dots\dots\dots\dots\dots\dots\dots\dots\dots\dots\dots\dots\dots$$

$$D_{109} = \dots\dots\dots\dots\dots\dots\dots\dots\dots\dots\dots\dots\dots\dots\dots\dots\dots\dots\dots$$

Step 2
Calculate the D$_{100}$ and D$_{115}$ values:

$$D_{100} = \dots\dots\dots\dots\dots\dots\dots\dots\dots\dots\dots\dots\dots\dots\dots\dots\dots\dots$$

$$D_{115} = \dots\dots\dots\dots\dots\dots\dots\dots\dots\dots\dots\dots\dots\dots\dots\dots$$

Step 3
Calculate the vitamin loss at 100 °C and 115 °C:

$$\left(\frac{C}{C_o}\right)_{100°C} = \dots\dots\dots\dots\dots\dots\dots\dots\dots\dots\dots\dots\dots\dots\dots\dots\dots\dots\dots$$

$$\left(\frac{C}{C_o}\right)_{115°C} = \dots\dots\dots\dots\dots\dots\dots\dots\dots\dots\dots\dots\dots\dots\dots\dots\dots$$

Exercise 14.5

For an initial spore load equal to 20 spores per can and a food spoilage microorganism with $D_{121} = 1$ min, calculate the spoilage probability if an F equivalent to $F_{121} = 6$ min was applied.

Solution

Calculate N from eqn (14.2):

$$N = \dots\dots\dots\dots\dots\dots\dots\dots\dots\dots\dots\dots\dots\dots\dots\dots$$

Therefore, the spoilage probability is:

$$\dots\dots\dots\dots\dots\dots\dots\dots\dots\dots\dots\dots\dots\dots\dots\dots\dots\dots\dots\dots\dots\dots\dots\dots\dots\dots\dots\dots\dots\dots$$

or 1 can in .... cans.

Exercise 14.6

A low acid solid food product was sterilized in a can with 76 mm diameter and 112 mm height. The autoclave temperature varied with time as given in the table below. An $F_o = 3$ min has been established in the laboratory as a minimum process sterilizing value for the slowest heating point of the can in order for this product to be safe. Determine if the applied thermal treatment is adequate to deliver an equivalent lethality of $F_o = 3$ min. It is given that the thermal diffusivity and the initial temperature of the product are 1.3×10^{-7} m^2/s and 30 °C respectively. If this thermal treatment is not enough, decide when the cooling period should start so that the lethality achieved is at least 3 min. Study the effect of steam temperature and cooling water temperature on the F value.

Time min	Temperature °C	Time min	Temperature °C	Time min	Temperature °C
0	40	31	125	62	40
1	60	32	125	63	40
2	90	33	125	64	40
3	125	34	125	65	40
4	125	35	125	66	40
5	125	36	125	67	40
6	125	37	125	68	40
7	125	38	125	69	40
8	125	39	125	70	40
9	125	40	125	71	40
10	125	41	83	72	40
11	125	42	56	73	40
12	125	43	40	74	40
13	125	44	40	75	40
14	125	45	40	76	40
15	125	46	40	77	40
16	125	47	40	78	40
17	125	48	40	79	40
18	125	49	40	80	40
19	125	50	40	81	40
20	125	51	40	82	40
21	125	52	40	83	40
22	125	53	40	84	40
23	125	54	40	85	40
24	125	55	40	86	40
25	125	56	40	87	40
26	125	57	40	88	40
27	125	58	40	89	40
28	125	59	40	90	40
29	125	60	40	91	40
30	125	61	40	92	40

Solution

Step 1
State your assumptions:

....

Step 2
The temperature in the center of the can vs. time will be calculated using the partial differential equation for a finite cylinder, eqn (14.5). The explicit finite differences method will be used for the solution of eqn (14.5) since there is not an analytical solution for the conditions applied:

$$\frac{\partial T}{\partial r} = \alpha \left(\frac{\partial^2 T}{\partial r^2} + \frac{1}{r}\frac{\partial T}{\partial r} + \frac{\partial^2 T}{\partial y^2} \right) \qquad (14.5)$$

If the temperature in the center is known, the lethality of the process for the time-temperature history of the geometric center of the can (the slowest heating point) can be calculated with eqn (14.6):

$$F = \int_0^t 10^{\frac{T-T_R}{z}} \, dt \approx \sum_0^t 10^{\frac{T-T_R}{z}} \Delta t \qquad (14.6)$$

The values $T_R = 121.1\,°C$ and $z = 10\,°C$ will be used, since the F_o value is given.

Step 3
The solution outlined in step 2 is shown in the spreadsheet program *Can sterilization.xls*.
To solve the problem, read the instructions, insert the input data, and iterate until a constant F value is reached. Observe in the plot that the maximum temperature at the center is reached after the onset of the cooling period. Explain this.

Step 4
If the F calculated is lower than 3 min (F_o given), increase the value of t for the "Onset of cooling period" to delay the beginning of cooling. This change will increase the heating period. Rerun the program (turn the SWITCH OFF, change the value of t, turn the SWITCH ON and iterate). Repeat step 3 until $F \geq 3$ min. Observe the sensitivity of F to t.

Step 5
Use the initial time for the "onset of cooling period" (t = 40 min). Change the steam temperature and rerun the program until $F \geq 3$ min. Observe the sensitivity of F to the steam temperature. Observe that low temperatures contribute little to sterilization (observe the minimum temperature at the center which gives an appreciable change in F).

Step 6

Use the initial values for the steam temperature and the time for the onset of cooling. Change the cooling water temperature and rerun the program until $F \geq 3$ min. Observe the sensitivity of F to the cooling water temperature. What problem might arise if the cooling water temperature is too low?

Exercise 14.7

Use your own time-temperature data for the autoclave and re-solve the previous problem.

Solution

Insert a sheet in the spreadsheet *Can sterilization.xls* (On the **Insert menu** select **Sheet**).
Write your own time-temperature values in the new sheet.
Change cell H10 accordingly to read the new data.
Run the program.

Exercise 14.8

For the destruction of microorganisms in whole milk that will be used as a substrate for yoghurt starter manufacture in a dairy, the milk is heated to 95 °C in an agitated jacketed tank, held at this temperature for 15 min, and then cooled to 45 °C before inoculation. If the mass of milk is 1000 kg, the heat transfer area in the jacket is 4 m², the overall heat transfer coefficient between heating medium and milk is 300 W/m² °C, the heat capacity of milk is 3.9 kJ/kg °C, and the heating medium temperature is 130 °C, calculate the number of log cycles reduction of a mesophilic microorganism with $D_{121} = 1$ min and of a thermophilic microorganism with $D_{121} = 6$ min. Assume that z is 10 °C for both microorganisms and neglect the lethal effect of the cooling period.

Solution

Step 1
Calculate the change of temperature of the milk with time during heating (remember that heating of an agitated liquid can be treated as heating of a body with negligible internal resistance).

Step 2
Insert the data in the spreadsheet *General method.xls*. Run the program and calculate F.

Step 3
Convert F to log cycles reduction using D.

Exercise 14.9

Assume that the acceptable level of spoilage risk for a canned food is 1 can in 100000 (spoilage probability of 10^{-5}). Calculate the required F value if the initial microbiological load is 50 spores per can with $D_{121} = 1$ min.

Exercise 14.10

Beer is pasteurized at 71 °C for 15 s or 74 °C for 10 s. Calculate z.

Exercise 14.11

The minimum pasteurization requirement in the USA for liquid eggs according to Ref. 14 is 60 °C for 3.5 min, while in Great Britain it is 64.4 °C for 2.5 min. Assuming that these treatments are equivalent, calculate the required time for an equivalent treatment in Denmark, where the pasteurization temperature could be up to 69 °C.

Chapter 15
Cooling and Freezing

Review Questions

Which of the following statements are true and which are false?

1. Ammonia (R 717) as a refrigerant in industrial refrigeration systems has a wide range of evaporating and condensing temperatures, but its vapors are toxic and form flammable mixtures with air.
2. Tubing in ammonia refrigeration units is made of copper.
3. A disadvantage of ammonia as a refrigerant is its low value of latent heat of vaporization.
4. The higher the value of the latent heat of vaporization of a refrigerant the lower the required refrigerant flow rate for a given refrigeration load.
5. Refrigerants containing chlorine may reach the upper atmosphere, react with ozone, and deplete the ozone layer.
6. Hydrofluorocarbons (HFCs) are replacing chlorofluorocarbons (CFCs) as refrigerants.
7. HFC-134a is used in refrigeration systems with moderately low temperatures.
8. Refrigerant 404a is used in refrigeration systems with low temperature applications.
9. Compressor, evaporator, condenser, and expansion valve are the main components of a mechanical refrigeration system.
10. Expansion of the refrigerant in the expansion valve of a Carnot cycle takes place at constant entropy.
11. Compression of the refrigerant in the compressor takes place at constant enthalpy.
12. Evaporation of the refrigerant in the evaporator takes place at constant pressure.
13. The condenser is at the low pressure side of a mechanical refrigeration system.
14. The latent heat of vaporization of the refrigerant is absorbed from the surroundings of the evaporator.
15. Refrigerant vapors leaving the evaporator may be superheated.
16. Liquid refrigerant leaving the condenser may be subcooled.

17. The coefficient of performance (COP) is equal to the ratio of the refrigeration effect to the net work input.
18. COP is always less than one.
19. The higher the temperature difference between condenser and evaporator, the higher the COP.
20. Refrigeration load includes heat transferred into the refrigeration space by conduction through the walls and by air entering into the space, heat removed from the product, heat generated by the product, heat generated by lights, motors, and people
21. Water evaporation from the surface of the product can result in considerable weight loss in blast air freezers.
22. Individually quick frozen (IQF) products can be produced in belt freezers and fluidized bed freezers.
23. Cryogenic freezing results in frozen products with large ice crystals.
24. The rate of freezing in plate freezers is high.
25. Water activity of a frozen product is only a function of temperature.
26. Bound water does not freeze.
27. Bound water affects water activity of a frozen product.
28. As water in a solution changes to ice, solute concentration increases and freezing point decreases.
29. Heat capacity and thermal conductivity below the freezing point change significantly with temperature.
30. Thawing is slower than freezing because the thermal conductivity of ice is lower than that of water.

Examples

Example 15.1

Lettuce was vacuum-cooled and then loaded into a refrigerated car for shipment to a market. The trip will last 48 h. Calculate the total amount of heat removed and the peak refrigeration load (maximum heat removal rate) if: the amount of lettuce loaded is 3000 kg, the temperature in the truck is 2 °C, the temperature of lettuce when loaded is 5 °C and will drop to 2 °C in 2 h, the heat capacity of lettuce is 4.02 kJ/kg °C, the area of the walls of the truck is 80 m^2, the overall heat transfer coefficient for the walls is 0.3 W/m2 °C, the outside air temperature is 20 °C, and the heat of respiration of lettuce in the temperature range of 2 to 5 °C is 35x10^{-3} W/kg.

Solution

Step 1
State your assumptions:
- The truck is initially at 2 °C.
- The truck is tightly closed so that air leakages are negligible.
- The cooling requirements of the packaging material are neglected.

Step 2
Calculate the cooling load:

i) sensible heat that must be removed to cool the product from 5 °C to 2 °C:

$$q_1 = mc_p(T_i - T_f) = (3000 \, kg) \, (4.02 \, kJ/kg \, °C) \, (5 - 2) \, °C = 36180 \, kJ$$

ii) heat of respiration:

$$q_2 = mq_R = (3000 \, kg) \left(35 \times 10^{-3} \frac{J}{s \, kg}\right) \left(\frac{1 \, kJ}{1000 \, J}\right) (48 \, h) \left(\frac{3600 \, s}{h}\right) = 18144 \, kJ$$

iii) heat losses:

$$q_3 = AU(T_a - T_f) \, t$$
$$= (80 \, m^2) \left(0.3 \frac{W}{m^2 \, °C}\right) \left(\frac{1 kJ}{1000 \, J}\right) (20 - 2 \, °C) \, (48 \, h) \left(\frac{3600 \, s}{h}\right) = 74650 \, kJ$$

Step 3
Calculate the total amount of heat removed:

$$q = q_1 + q_2 + q_3 = 36180 \, kJ + 18144 \, kJ + 74650 \, kJ = 128974 kJ$$

Step 4
Calculate the peak refrigeration load:

i) Sensible heat:

$$q_{1p} = \frac{q_1}{t} = \frac{36180 \, kJ}{2*3600 \, s} = 5.025 \, kW$$

ii) Heat of respiration:

$$q_{2p} = (3000 \, kg) \left(35 \times 10^{-3} \frac{W}{kg}\right) = 105 \, W = 0.105 \, kW$$

iii) Heat losses:

$$q_{3p} = AU(T_a - T_f) = (80 \, m^2) \left(0.3 \frac{W}{m^2 \, °C}\right) (20 - 2 \, °C) = 432 \, W = 0.432 \, kW$$

iv) Peak refrigeration load:

$$q_p = q_{1p} + q_{2p} + q_{3p} = 5.025 + 0.105 + 0.432 = 5.562 \, kW$$

Example 15.2

The cooling in Example 15.1 is provided by a mechanical refrigeration system using HFC-134a refrigerant with the evaporator temperature being −5 °C and the condenser temperature 40 °C. Plot a diagram of the refrigeration system, present the conditions at the main points of the system on a pressure-enthalpy chart and a temperature-entropy chart, and calculate the COP. Assume the refrigeration unit operates at saturation conditions. Neglect the heat load from fans and other components.

Solution

Step1
Draw the diagram of the refrigeration system:

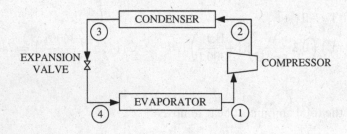

Step 2
Locate the points on the pressure-enthalpy and the temperature-entropy charts:

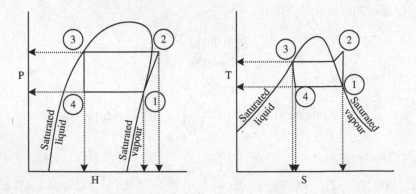

Step 3
Read pressure, enthalpy and entropy from the pressure-enthalpy chart and temperature-entropy chart of refrigerant HFC-134a or from a table with pressure, enthalpy and entropy values for HFC-134a:

Point 1: $P_1 = 243$ kPa, H_1 satur.vapor $= 396$ kJ/kg, S_1 satur.vapor $= 1.73$ kJ/kgK

Point 2: $P_2 = 1018$ kPa, H_2 superheat.vapor $= 426$ kJ/kg, S_2 superheat.vapor
        $= 1.73$ kJ/kg

Point 3: $P_3 = 1018$ kPa, H_3 satur.liquid $= 257$ kJ/kg, S_3 satur.liquid $= 1.19$ kJ/kgK

Point 4: $P_4 = 243$ kPa, H_4 liquid+vapor $= 257$ kJ/kg, S_4 liquid+vapor $= 1.21$ kJ/kgK

Step 4
Write an enthalpy balance on the evaporator:

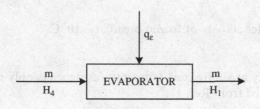

$$mH_4 + q_e = mH_1 \tag{15.1}$$

Solve eqn (15.1) for q_e and find the rate of heat absorbed in the evaporator:

$$q_e = m(H_1 - H_4) \tag{15.2}$$

Step 5
Write an enthalpy balance on the compressor:

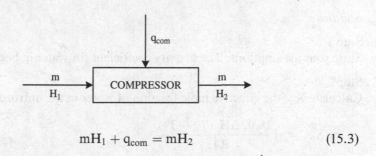

$$mH_1 + q_{com} = mH_2 \tag{15.3}$$

Step 6
Solve eqn (15.3) for q_{com} and find the rate of work done by the compressor on
the refrigerant:

$$q_{com} = m(H_2 - H_1) \tag{15.4}$$

Step 7
Calculate the COP:

$$COP = \frac{q_e}{q_{com}} = \frac{m(H_1 - H_4)}{m(H_2 - H_1)} = \frac{H_1 - H_4}{H_2 - H_1} = \frac{396 - 257}{426 - 396} = 4.6$$

Comment: The calculated COP value is higher than in practice due to the ideal conditions assumed. Actual COP values are usually in the range of 1 to 2.

Exercise 15.3

Calculate the water activity of frozen bread at $-10\,°C$.

Solution

The water activity at subfreezing temperatures depends only on temperature. It can be calculated from Ref. 11:

$$\ln(a_w) = 0.00969T = 0.00969^*(-10) = -0.0969$$

which gives:

$$a_w = 0.908$$

Example 15.4

Calculate the effective molecular weight of solutes, as well as the weight fraction of bound water, of freezable water, and of ice, for lean sirloin beef at $-20\,°C$ if the moisture content of the unfrozen beef is 71.7% and the initial freezing point is $-1.7\,°C$.

Solution

Step 1
State your assumptions: The activity coefficient for water in beef is 1.

Step 2
Calculate X_w, the effective mole fraction of water in the unfrozen food:

$$\ln(X_w) = -\frac{18.02\Delta H_o(T_o - T)}{RT_o^2}$$

$$= -\frac{(18.02\,kg/kmol)\,(333.6\,kJ/kg)\,(273 - 271.3)°C}{(8.314\,kJ/kmol\,K)\,(273\,K)^2} = -0.0165$$

which gives

$$X_w = 0.9836$$

Step 3
Calculate the effective molecular weight of solutes in the food, M_s.
The mole fraction of water in the food is:

$$X_w = \frac{\dfrac{x_{wo}}{18.02}}{\dfrac{x_{wo}}{18.02} + \dfrac{x_s}{M_s}}$$

Solve for M_s:

$$M_s = \frac{18.02 X_w (1 - x_{wo})}{x_{wo}(1 - X_w)} = \frac{18.02(0.9836)(1 - 0.717)}{0.717(1 - 0.9836)} = 426.6$$

where x_{wo} is the weight fraction of water in the food and
x_s is the weight fraction of solids in the food ($x_s = 1 - x_{wo}$).

Step 4
Calculate the weight fraction of bound water, x_{wb}:
Assume b = 0.32 kg water bound/kg of solute (typical values for the constant b are given in Ref. 10):

$$x_{wb} = b^* x_s = b(1 - x_{wo}) = 0.32\,(1 - 0.717) = 0.091$$

or 9.1% of food is bound water non-freezable.

Step 5
Calculate the weight fraction of freezable water, x_{wf}:

$$x_{wf} = x_{wo} - x_{wb} = 0.717 - 0.091 = 0.626$$

or 62.6% of food is freezable water.

Step 6
Calculate the weight fraction of ice in the food, x_I (Ref. 11):

$$x_I = (x_{wo} - Bx_s)\left(\frac{T_{if} - T}{T_o - T}\right)$$

where
$$B = b - 0.5\frac{M_w}{M_s} = 0.32 - 0.5\frac{18.02}{426.6} = 0.299 \text{ kg water bound/kg of solute}$$

and
$$Bx_s = 0.299^*0.283 = 0.085 \text{ kg water bound/kg food}$$

$$x_I = (x_{wo} - Bx_s)\left(\frac{T_{if} - T}{T_o - T}\right) = (0.717 - 0.085)\left(\frac{-1.7 - (-20)}{0 - (-20)}\right) = 0.578$$

or 57.9% of food at $-20\,°C$ is ice.

Example 15.5

Calculate the refrigeration load when 100 kg of lean sirloin beef is frozen from 0 °C to −20 °C. It is given that the moisture content of the unfrozen beef is 71.7%, the initial freezing point is −1.7 °C, the heat capacity of the unfrozen beef is 3.08 kJ/kg °C, and the heat capacity of the fully frozen beef is 1.5 kJ/kg °C, of the solids 1.25 kJ/kg °C and of the ice 2.0 kJ/kg °C.

Solution

As calculated in the previous example $Bx_s = 0.085$

1st solution

Step 1
Calculate the enthalpy of beef at 0 °C using Schwartzberg's equation (Ref. 11) with −40 °C as reference temperature.
The enthalpy will include the sensible heat above the initial freezing point (H_s) and the enthalpy at the initial freezing point H_{Tif}:

$$H = H_s + H_{Tif} = c_{pu}(T - T_{if})$$

$$+ \left[c_f + (x_{wo} - Bx_s)\left(\frac{\Delta H_o}{T_o - T_R} \frac{T_o - T_{if}}{T_o - T_{if}} \right) \right](T_{if} - T_R) =$$

$$= 3.08(0 - (-1.7))$$

$$+ \left[(1.5 + (0.717 - 0.085)\left(\frac{333.6}{0 - (-40)} \right) \right](-1.7 - (-40)) =$$

$$= 264.6 \text{ kJ/kg}$$

Step 2
Calculate the enthalpy at −20 °C using Schwartzberg's equation:

$$H = \left[c_f + (x_{wo} - Bx_s)\left(\frac{\Delta H_o}{T_o - T_R} \frac{T_o - T_{if}}{T_o - T} \right) \right](T - T_R) =$$

$$= \left[(1.5 + (0.717 - 0.085)\left(\frac{333.6}{0 - (-40)} \frac{0 - (-1.7)}{0 - (-20)} \right) \right](-20 - (-40)) =$$

$$= 39.0 \text{ kJ/kg}$$

Step 3
Calculate the heat that must be removed:

$$q = m\Delta H = (100 \text{ kg})(264.6 - 39.0) \text{ kJ/kg} = 22560 \text{ kJ}$$

2nd solution
Read the enthalpy values from Riedel's chart for beef (as given in Ref. 9).
Step 1
Read the enthalpy value at 0 °C on the y-axis for 71.7% water content (x-axis)
and 0 °C temperature (0 °C curve)

$$H_{0\,°C} = 283.8 \text{ kJ/kg}$$

Step 2
Read the enthalpy value at -20 °C on the y-axis for 71.7% water content (x-axis)
and-20 °C temperature (-20 °C curve):

$$H_{-20\,°C} = 38.2 \text{ kJ/kg}$$

Step 3
Calculate the heat that must be removed:

$$q = m\Delta H = (100 \text{ kg}) \ (283.8 - 38.2) \text{ kJ/kg} = 24560 \text{ kJ}$$

Example 15.6

Peas are IQF frozen in a fluidized bed using air at $-35\,°C$ as a freezing medium.
Calculate the time necessary for the temperature at the center of a pea to drop to
$-20\,°C$ if the initial temperature of the peas is 15 °C, the initial freezing point is
$-0.6\,°C$, the diameter of a pea is 8 mm, the heat transfer coefficient is 150 W/
m2 °C, the moisture content is 79%, the thermal conductivity of the unfrozen
product is 0.315 W/m °C, the thermal conductivity of the frozen product is
0.48 W/m °C, the heat capacity of the unfrozen product is 3.31 kJ/kg °C, the
heat capacity of the frozen product is 1.76 kJ/kg °C, the density of the unfrozen
product is 1032 kg/m³, the density of the frozen product is 970 kg/m³, and the
protein content of the peas is 5.42%.

1st solution

Use Plank's equation to calculate the time necessary to remove the latent heat of
freezing and the solution of Fick's 2nd law to calculate the time for the sensible
heat removal above and below the freezing point.

Step 1
Calculate the time necessary to remove the sensible heat and bring the mass
average temperature of a pea down to the initial freezing point.

Use the spreadsheet program *Heat Transfer-Internal and External Resistan-
ce.xls*. Insert parameter values in the sheet "Sphere" and iterate until the
average temperature is equal to $-0.6\,°C$. Read the time:

$$t_1 = 13.3s$$

Comment: This is an approximate method since freezing occurs near the surface of the pea by the time its mean temperature reaches the initial freezing point.

Step 2

Calculate the time necessary to remove the latent heat using Plank's equation:

$$t_2 = \frac{\rho \lambda}{T_{if} - T_m} \left(\frac{P\alpha}{h} + \frac{R\alpha^2}{k_f} \right)$$

For a sphere, $P = 1/6$, $R = 1/24$ and $\alpha = D$

Substitute values and calculate t_2:

$$t_2 = \frac{\rho \lambda}{T_{if} - T_m} \left(\frac{P\alpha}{h} + \frac{R\alpha^2}{k_f} \right) =$$

$$= \frac{\left(970 \,\text{kg/m}^3\right) (0.79{*}333600 \,\text{J/kg})}{-0.6 - (-35)} \left(\frac{(1/6)\,(0.008\,\text{m})}{150\,\text{W/m2}^\circ\text{C}} + \frac{(1/24)\,(0.008\,\text{m})^2}{0.48\,\text{W/m}^\circ\text{C}} \right)$$

$$= 107.3\,\text{s}$$

Comment: This solution is an approximate solution due to the limiting assumptions of Plank's equation.

Step 3

Calculate the time to cool the frozen product from the initial freezing point to the final temperature. Assume that only sensible heat is removed.

Use the spreadsheet program *Heat Transfer-Internal and External Resistance.xls*. Insert parameter values in the sheet "Sphere" and iterate until the temperature at the center is equal to $-20\,^\circ\text{C}$. Read the time:

$$t_3 = 21.6\text{s}$$

Comment: This solution is an approximate solution because a) the pea temperature is non uniform at the start of t_3, and b) freezing of water is continuous as the temperature drops below the initial freezing point.

Step 4

Calculate the total time:

$$t = t_1 + t_2 + t_3 = 13.3 + 107.3 + 21.6 = 142.2\,\text{s}$$

2nd solution

Use the Cleland and Earle modification of Plank's equation, eqn (15.5), to calculate the freezing time. Eqn (15.5) takes into account sensible heat and latent heat removal:

$$t = \frac{\Delta H_{10}}{T_{if} - T_m} \left(\frac{P\alpha}{h} + \frac{R\alpha^2}{k_f} \right) \left[1 - \frac{1.65 \, \text{Ste}}{k_f} \ln \left(\frac{T_c - T_m}{-10 - T_m} \right) \right] \quad (15.5)$$

where ΔH_{10} is the volumetric enthalpy change of the food between the initial freezing temperature and the final center temperature (assumed to be $-10\,°C$).

P and R are now functions of Ste (Stefan number) and Pk (Plank number):

$$Pk = \frac{c_u(T_o - T_{if})}{\Delta H_{10}}$$

$$\text{Ste} = \frac{c_f(T_{if} - T_m)}{\Delta H_{10}}$$

Therefore it is necessary to calculate ΔH_{10}, Pk, Ste, P, R, and finally t.

Step 1
Calculate the volumetric enthalpy change ΔH_{10}.

1) Calculate H at the initial freezing point using Schwartzberg's equation.

i) Calculate the bound water.
The bound water can be calculated as in Example 15.4 or alternatively by using the relationship (Ref. 8):

$$x_{wb} = 0.4x_p$$

where x_p is the protein content of the food. In this case,

$$x_{wb} = 0.4 {}^* x_p = 0.4 {}^* 0.0542 = 0.0217$$

Comment: The bound water content calculated with this method is probably too low.

ii) Calculate H:

$$H = \left[c_f + (x_{wo} - bx_s) \left(\frac{\Delta H_o}{T_o - T_R} \frac{T_o - T_{if}}{T_o - T} \right) \right] (T - T_R) =$$

$$= \left[(1.76 + (0.79 - 0.0217) \left(\frac{333.6}{0 - (-40)} \frac{0 - (-0.6)}{0 - (-0.6)} \right) \right] (-0.6 - (-40)) =$$

$$= 321.8 \, kJ/kg$$

2) Calculate H at $-10°C$ using Schwartzberg's equation:

$$H = \left[c_f + (x_{wo} - bx_s) \left(\frac{\Delta H_o}{T_o - T_R} \frac{T_{o-}T_{if}}{T_o - T} \right) \right] (T - T_R) =$$

$$= \left[(1.76 + (0.79 - 0.0217) \left(\frac{333.6}{0 - (-40)} \frac{0 - (-0.6)}{0 - (-10)} \right) \right] (-10 - (-40)) =$$

$$= 64.3 \, \text{kJ/kg}$$

3) Therefore, the enthalpy difference between $0°C$ and $-10°C$ is:

$$\Delta H = 321.8 - 64.3 = 257.5 \text{ kJ/kg}$$

or the volumetric enthalpy difference is:

$$\Delta H_{10} = \rho_f \Delta H = \left(970 \, \text{kg/m}^3 \right) (257.5 \text{ kJ/kg}) = 249775 \, \text{kJ/m}^3$$

Step 2
Calculate Plank, Stefan and Biot numbers.

i) The volumetric heat capacity of the unfrozen product is:

$$c_u = \rho_u c_u = \left(1032 \, \text{kg/m}^3 \right) (3.31 \, \text{kJ/kg°C}) = 3415.9 \, \text{kJ/m3°C}$$

ii) The volumetric heat capacity of the frozen product is:

$$c_f = \rho_f c_f = \left(970 \, \text{kg/m}^3 \right) (1.76 \, \text{kJ/kg°C}) = 1707.2 \, \text{kJ/m3°C}$$

iii) The Plank number is:

$$Pk = \frac{c_u(T_o - T_{if})}{\Delta H_{10}} = \frac{(3415.9 \text{ kJ/m}^3 \, °C) \, (15 - (-0.6))°C}{249775 \text{ kJ/m}^3} = 0.213$$

iv) The Stefan number is:

$$Ste = \frac{c_f(T_{if} - T_m)}{\Delta H_{10}} = \frac{(1707.2 \text{ kJ/m}^3 \, °C) \, (-0.6 - (-35))°C}{249775 \text{ kJ/m}^3} = 0.235$$

v) The Biot number is:

$$Bi = \frac{h(D/2)}{k_f} = \frac{(150\,\text{W/m2°C})\,(0.004\,\text{m})}{0.48\ \text{W/m°C}} = 1.25$$

Step 3
Calculate P and R from expressions for a sphere (see, e.g., Ref. 8):

$$P = 0.1084 + 0.0924\,\text{Pk} + \text{Ste}\left(0.231\,\text{Pk} - \frac{0.1557}{Bi} + 0.6739\right) =$$

$$= 0.1084 + 0.0924*0.213 + 0.235\left(0.231*0.213 - \frac{0.1557}{1.25} + 0.6739\right)$$

$$= 0.269$$

$$R = 0.0784 + \text{Stc}(0.0386\,\text{Pk} - 0.1694)$$
$$= 0.0784 + 0.235\,(0.0386*0.213 - 0.1694)$$
$$= 0.041$$

Step 4
Substitute values in eqn (15.5) and calculate the time:
$$t = \frac{\Delta H_{10}}{T_{if} - T_m}\left(\frac{PD}{h} + \frac{RD^2}{k_f}\right)\left[1 - \frac{1.65\,\text{Ste}}{k_f}\ln\left(\frac{T_c - T_m}{-10 - T_m}\right)\right] =$$

$$\frac{249775000\,\text{J/m}^3}{-0.6 - (-35)°C}\left(\frac{(0.269)\,(0.008\,\text{m})}{150\,\text{W/m2°C}} + \frac{(0.041)\,(0.008\,\text{m})^2}{0.48\,\text{W/m°C}}\right)*$$

$$*\left[1 - \frac{1.65*0.235}{0.48\,\text{W/m°C}}\ln\left(\frac{-20 - (-35)}{-10 - (-35)}\right)\right] = 203.2\ \text{s}$$

3rd solution

Use the modification of Plank's equation proposed by Pham to calculate the freezing time for a slab, taking into account sensible heat and latent heat removal, and then divide by the shape factor E to calculate the freezing time for the sphere:

$$t = \frac{1}{E}\left(\frac{\Delta H_1}{\Delta T_1} + \frac{\Delta H_2}{\Delta T_2}\right)\left(\frac{P\alpha}{h} + \frac{R\alpha^2}{k_f}\right) \tag{15.6}$$

with $P = 1/2, R = 1/8$, and α = characteristic dimension:

$$\Delta H_1 = c_u(T_o - T_{fm})$$

$$\Delta H_2 = \lambda + c_f(T_{fm} - T_c)$$

c_u and c_f volumetric heat capacities and λ volumetric latent heat of freezing:

$$\Delta T_1 = \frac{T_o + T_{fm}}{2} - T_m$$

$$\Delta T_2 = T_{fm} - T_m$$

$$T_{fm} = 1.8 + 0.263T_c + 0.105T_m$$

$$E = 1 + \frac{1 + \frac{2}{Bi}}{\beta_1^2 + \frac{2\beta_1}{Bi}} + \frac{1 + \frac{2}{Bi}}{\beta_2^2 + \frac{2\beta_2}{Bi}}$$

$$\beta_1 = \frac{A}{\pi R^2}$$

$$\beta_2 = \frac{V}{\frac{4}{3}\pi R^3 \beta_1}$$

Therefore it is necessary to calculate $\Delta H_1, \Delta H_2, \Delta T_1, \Delta T_2, T_{fm}, E$, and finally t.

Step 1

1 Calculate ΔH_1 and ΔH_2:

i) The volumetric heat capacity of the unfrozen product is:

$$c_u = \rho_u c_u = (1032\,kg/m^3)\,(3.31\,kJ/kg°C) = 3415.9\,kJ/m\,3°C$$

ii) The volumetric heat capacity of the frozen product is:

$$c_f = \rho_f\,c_f = (970\,kg/m^3)\,(1.76\,kJ/kg°C) = 1707.2\,kJ/m3°C$$

iii) The volumetric latent heat of freezing is:

$$\lambda = (x_{wo} - x_{wb})\rho\,\lambda_w = (0.79 - 0.0217)\,(1032\,kg/m^3)\,(333.6\,kJ/kg)$$
$$= 264507\,kJ/m^3$$

iv) The mean freezing temperature is:

$$T_{fm} = 1.8 + 0.263 T_c + 0.105 \, T_m$$

$$= 1.8 + 0.263 \, (-20) + 0.105 \, (-35) = -7.1$$

Therefore,

$$\Delta H_1 = c_u(T_o - T_{fm}) = (3415.9 \text{ kJ/m3°C}) \, (15 - (-7.1)) \, ^\circ C = 75491 \, \text{kJ/m}^3$$

$$\Delta H_2 = \lambda + c_f(T_{fm} - T_c) = 264507 \text{ kJ/m}^3$$
$$+ (1707.2 \text{ kJ/m3°C}) \, (-7.1 - (-20)) \, ^\circ C =$$
$$= 286530 \, \text{kJ/m}^3$$

Step 2
Calculate Δ_{T1} and Δ_{T2}:

$$\Delta T_1 = \frac{T_o + T_{fm}}{2} - T_m = \frac{15 + (-7.1)}{2} - (-35) = 38.9 \, ^\circ C$$

$$\Delta T_2 = T_{fm} - T_m = -7.1 - (-35) = 27.9 \, ^\circ C$$

Step 3
Calculate E:

$$\beta_1 = \frac{A}{\pi R^2} = \frac{\pi R^2}{\pi R^2} = 1$$

$$\beta_2 = \frac{V}{\frac{4}{3}\pi R^3 \beta_1} = \frac{\frac{4}{3}\pi R^3}{\frac{4}{3}\pi R^3 * 1} = 1$$

$$E = 1 + \frac{1 + \frac{2}{Bi}}{\beta_1^2 + \frac{2\beta_1}{Bi}} + \frac{1 + \frac{2}{Bi}}{\beta_2^2 + \frac{2\beta_2}{Bi}} = 1 + \frac{1 + \frac{2}{Bi}}{1 + \frac{2}{Bi}} + \frac{1 + \frac{2}{Bi}}{1 + \frac{2}{Bi}} = 1 + 1 + 1 = 3$$

Step 4
Substitute values in eqn (15.6) and calculate t:

$$t = \frac{1}{E}\left(\frac{\Delta H_1}{\Delta T_1} + \frac{\Delta H_2}{\Delta T_2}\right)\left(\frac{D}{2h} + \frac{D^2}{8k_f}\right) =$$

$$= \frac{1}{3}\left(\frac{75491000\,\text{J/m}^3}{38.9°\text{C}} + \frac{286530000\,\text{J/m}^3}{27.9°\text{C}}\right)$$

$$\left(\frac{0.008\,\text{m}}{2(150\,\text{W/m}^2°\text{C})} + \frac{(0.008\,\text{m})^2}{8\,(0.48\,\text{W/m}°\text{C})}\right) =$$

$$= 176.4\,\text{s}$$

Comment: The 2nd and 3rd solutions give comparable results, while the 1st solution gives a substantially lower freezing time. It was expected that the 1st solution will not give an accurate prediction due to the imposed simplifying assumptions.

Exercises

Exercise 15.1

1000 kg of meat will be cooled from 25 °C to 5 °C in 2 h. Calculate the rate of heat removal if the heat capacity of the meat is 3.35 kJ/kg °C. Assume that the effect of water evaporated from the surface of the meat is negligible.

Solution

Step 1
Calculate the sensible heat that must be removed:

$$q_s = \dots\dots\dots\dots\dots\dots\dots\dots\dots\dots\dots\dots$$

Step 2
Calculate the rate of heat removal:

$$q = \dots\dots\dots\dots\dots\dots\dots\dots\dots\dots\dots\dots$$

Exercise 15.2

1000 kg of unshelled green beans will be cooled from 25 °C to 5°C in 2 h. Calculate the rate of heat removal if the heat capacity of beans is 3.35 kJ/kg °C and the average heat of respiration between 5 and 25 °C is 0.2 W/kg. Assume that the effect of water evaporated from the surface of the beans is negligible.

Solution

Step 1
Calculate the sensible heat that must be removed:

$$q_s = \dots\dots\dots\dots\dots\dots\dots\dots\dots\dots\dots\dots$$

Step 2
Calculate the heat of respiration:

$$q_R = \dots\dots\dots\dots\dots\dots\dots\dots\dots\dots\dots$$

Step 3
Calculate the total heat:

$$q = \dots\dots\dots\dots\dots\dots\dots\dots\dots\dots\dots$$

Step 4
Calculate the rate of heat removal:

$$q = \dots\dots\dots\dots\dots\dots\dots\dots\dots\dots\dots$$

Exercise 15.3

To rapidly cool white wine grape musts for premium quality white wine production, dry ice is added to the must immediately after crushing the grapes. Calculate the amount of dry ice per kg of must required to cool the must from 25 °C to 10 °C if the heat capacity of grapes is 3.6 kJ/kg °C, the enthalpy of dry ice is −31.9 kJ/kg and the enthalpy of CO_2 gas at 10 °C and 1 atm pressure is 380.9 kJ/kg.

Solution

Step 1
Draw the process diagram:

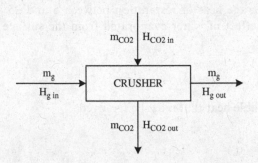

Step 2
State
assumptions:..

Step 3
Write an enthalpy balance for the process and calculate m_{CO2}:

$$m_{co_2} = ...$$

Exercise 15.4

Calculate the time required to cool a rectangular piece of ham with dimensions 0.1 m × 0.1 m × 0.3 m in a refrigerator from 60 °C to 5 °C if the air temperature in the refrigerator is 1 °C, the heat transfer coefficient in the refrigerator is 10 W/m² °C, the density of the ham is 1080 kg/m³, the thermal conductivity is 0.38 W/m °C, and the heat capacity is 3750 J/kg °C.

Solution

Step 1
Calculate the Biot number for the x, y, and z directions:

$$Bi_x = ...$$

$$Bi_y = ...$$

$$Bi_z = ...$$

Step 2
Use the spreadsheet program *Heat Transfer-Internal and External Resistance.xls*
or *Heat Transfer-Negligible External Resistance.xls* (depending on the Bi values).
Read the instructions, insert parameter values, and iterate until the mass
average temperature is equal to 5 °C. Read the time.

Exercise 15.5

Calculate the ice fraction and the unfrozen water fraction in salmon at −20 °C if
the initial freezing point is −2.2 °C, the moisture content of the unfrozen fish is
76.4%, and b = 0.25 kg water/kg solids. Test the sensitivity of the ice fraction
and the unfrozen water fraction on the value of the temperature and b.

Solution

Use the spreadsheet program *Ice fraction.xls* to calculate the ice fraction and the
unfrozen water fraction.
Run the program for different b values and different T values and see how the
ice fraction and the unfrozen water fraction are affected.

Exercise 15.6

Calculate the heat that is removed when 1000 kg cod is frozen from 5 °C to
−26 °C, the recommended temperature for frozen fish storage, if the moisture
content of the unfrozen fish is 80.3%, the initial freezing point is −2.2 °C, the
heat capacity of the unfrozen fish is 3.77 kJ/kg °C, the heat capacity of the fully
frozen fish is 2.0 kJ/kg °C, and 0.3 kg water are bound per kg solid.

Solution

Step 1
Calculate the enthalpy of fish at 5 °C with Schwartzberg's equation, using
−40 °C as reference temperature:

$$H = c_{pu}(T - T_{if}) + \left[c_f + (x_{wo} - bx_s)\left(\frac{\Delta H_o}{T_o - T_R}\right)\right](T_{if} - T_R) =$$

$$= \dots\dots\dots\dots\dots\dots\dots\dots\dots\dots\dots\dots\dots\dots\dots\dots\dots\dots$$

Step 2
Calculate the enthalpy of fish at -26 °C using Schwartzberg's equation:

$$H = \left[c_f + (x_{wo} - bx_s)\left(\frac{\Delta H_o}{T_o - T_R}\frac{T_o - T_{if}}{T_o - T}\right)\right](T - T_R) =$$

$$= \dots\dots\dots\dots\dots\dots\dots\dots\dots\dots\dots\dots\dots\dots\dots\dots\dots\dots$$

Step 3
Calculate the heat that must be removed:

$$q = \dots\dots\dots\dots\dots\dots\dots\dots\dots\dots\dots\dots\dots\dots\dots\dots\dots\dots\dots\dots$$

Exercise 15.7

Calculate the time required to freeze the cod of Exercise 15.6 if an air blast freezer is used with air temperature at $-30\,°C$. It is given that the heat transfer coefficient is $20\ W/m^2\,°C$, the density of the unfrozen fish is $1055\ kg/m^3$, that of the frozen fish is $990\ kg/m^3$, and the thermal conductivity of the frozen fish is 1.7 $W/m\,°C$. Assume that each cod has 2 kg weight, $60\ cm^2$ cross section, and 6 cm thickness in the thickest part of the body.

Solution

Use Pham's modification of Plank's equation to calculate the freezing time:

Step 1
Calculate ΔH_1 and ΔH_2:

   i) Calculate the mean freezing temperature:

$$T_{fm} = \dots\dots\dots\dots\dots\dots\dots\dots\dots\dots\dots\dots\dots\dots\dots\dots\dots\dots\dots\dots$$

   ii) Calculate ΔH_1 :

$$\Delta H_1 = \dots\dots\dots\dots\dots\dots\dots\dots\dots\dots\dots\dots\dots\dots\dots\dots\dots\dots\dots$$

   iii) Calculate the bound water x_{wb} using the spreadsheet program *Ice fraction.xls*:

$$x_{wb} = \dots\dots\dots\dots\dots\dots\dots\dots\dots\dots\dots\dots\dots$$

   iv) Calculate λ:

$$\lambda = \dots\dots\dots\dots\dots\dots\dots\dots\dots\dots\dots\dots\dots$$

   v) Calculate ΔH_2:

$$\Delta H_2 = \dots\dots\dots\dots\dots\dots\dots\dots\dots\dots\dots\dots\dots\dots\dots\dots\dots$$

Step 2
Calculate ΔT_1 and ΔT_2:

$$\Delta T_1 = \dots\dots\dots\dots\dots\dots\dots\dots\dots$$

$$\Delta T_2 = \dots\dots\dots\dots\dots\dots\dots\dots\dots$$

Step 3
Calculate E using R as the characteristic dimension and A as the cross section through the thickest part of the fish (Ref. 10):

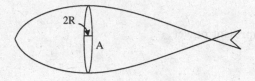

i) Calculate $\beta_1 = \dots\dots\dots\dots\dots\dots\dots\dots\dots$

ii) Calculate β_2 using V equal to the volume of the fish:

$$\beta_2 = \dots\dots\dots\dots\dots\dots\dots\dots\dots$$

iii) Calculate Bi:

$$Bi = \frac{hR}{k_f} = \dots\dots\dots\dots\dots\dots\dots\dots\dots$$

iv) Calculate the shape factor E:

$$E = \dots\dots\dots\dots\dots\dots\dots\dots\dots\dots\dots\dots\dots\dots\dots\dots\dots\dots\dots\dots\dots$$

Step 4
Calculate the time:

$$t = \dots\dots\dots\dots\dots\dots\dots\dots\dots\dots\dots\dots\dots\dots\dots\dots\dots\dots\dots\dots$$

Exercise 15.8

Calculate the required compressor power in Exercise 15.7 if the evaporator temperature is $-40\,°C$, the condenser temperature is $40\,°C$, the refrigeration unit uses refrigerant 404A at saturation conditions with compressor efficiency equal to 0.85 and the product cooling load represents 80% of the total heat removal load.

Solution

Step 1
Find the enthalpy of the refrigerant before and after the evaporator and the compressor from the pressure-enthalpy chart of refrigerant 404A or from a table with pressure enthalpy values for 404A.

Step 2
Calculate the heat absorption rate in the evaporator. Take into account that the product cooling load is 80% of the total load.

Step 3
Calculate the compressor power.

Exercise 15.9

Rectangular pieces of ground meat with dimensions 5 cm x 5 cm x 20 cm are placed in a batch freezer. The initial temperature of the meat is 5 °C, the initial freezing point is −2.2 °C, the moisture content is 73%, the bound water fraction is 0.05, the heat capacity is 2.9 kJ/kg °C and 1.4 kJ/kg °C for the unfrozen product and the fully frozen product respectively, the thermal conductivity is 0.5 W/m °C and 1.3 W/m °C for the unfrozen product and the fully frozen product respectively, the density is 1045 kg/m^3 and 1030 kg/m^3 for the unfrozen and the frozen product respectively. The heat transfer coefficient is equal to 10 W/m^2 °C. The air temperature in the freezer varies due to the ON-OFF operation of the compressor according to the relationship $T = A+B\sin(2\pi Nt)$ with $A = -20$ °C, $B = 2$ °C and $N = 2$ h^{-1}. Calculate and plot the temperature at the center of a piece of meat with time until it reaches −20 °C.

Solution

Since the length of the pieces of meat is much higher than the other two dimensions, it can be assumed that the heat transfer will be in two dimensions only. Use the spreadsheet program *Freezing temperature 2D.xls* to calculate and plot the temperature. Read the instructions; run the program until the temperature in the center is equal to −20 °C. Read the time necessary to freeze the product to −20 °C. Observe the freezing curve for the center and the corner.

Turn the SWITCH OFF; change the bound water value to 0.08; run the program again; compare the results with the previous case.

Turn the SWITCH OFF; change the heat transfer coefficient value to 30 W/ m^2 °C; run the program again; compare the time to reach −20 °C in the center with the previous case; compare the freezing curves; observe the temperature fluctuation in the product after the center temperature reaches −20 °C; compare it with the previous cases; explain the results.

Chapter 16
Evaporation

Review Questions

Which of the following statements are true and which are false?

1. Evaporation is used to decrease the weight and volume of a liquid product.
2. Concentration of a liquid food increases the shelf life of the food since it decreases the water activity.
3. Falling film evaporators are not suitable for heat sensitive products.
4. Agitated film evaporators are suitable for viscous liquids.
5. Fruit juices are concentrated in falling film evaporators.
6. Gelatin is concentrated in agitated film evaporators.
7. Milk is concentrated in short tube vertical evaporators.
8. Tomato juice is concentrated in forced-circulation evaporators.
9. Falling film tube evaporators are more compact than plate evaporators.
10. Natural circulation evaporators are suitable for heat sensitive liquid foods.
11. Natural circulation evaporators are not suitable for viscous liquids.
12. Forced circulation evaporators are suitable for viscous liquid foods.
13. Rising film evaporators require smaller temperature differences than falling film evaporators.
14. Rising film evaporators are short tube evaporators.
15. Falling film evaporators are long tube evaporators.
16. Backward feed evaporators are suitable for highly viscous liquid foods.
17. Backward feed evaporators are suitable when the fresh feed is cold.
18. Boiling under vacuum decreases the heat damage to heat sensitive products.
19. Multiple effect evaporators reuse the latent heat of vapor and thus save energy.
20. The overall temperature difference in a multiple effect evaporator will increase if the pressure in the last effect increases, while steam pressure in the first effect remains unchanged.
21. The concentration of the final product will increase if the overall temperature difference in an evaporator is increased.
22. If the cooling water temperature in the condenser is decreased, the final product concentration will increase.
23. If the feed flow rate is increased, the final product concentration will increase.

24. If the steam temperature is increased, the final product concentration will increase.
25. Dühring line plots are plots of the boiling point of a solution vs. the boiling point of pure water at the same pressure.
26. Boiling point elevation results in a decrease of the available temperature difference.
27. The hydrostatic head in rising film evaporators reduces the available temperature difference.
28. Residence time in a falling film evaporator is smaller than in a forced circulation evaporator.
29. Centrifugal evaporators have a residence time in the heating zone of a few seconds.
30. The residence time in a forced circulation evaporator is high.
31. The overall heat transfer coefficient decreases as the concentration of the product increases.
32. Overall heat transfer coefficients up to 2500 W/m^2K are usually encountered in the first effect of a falling film evaporator.
33. Fouling of the evaporator tubes reduces its capacity.
34. Preheating the feed to the boiling point of the first effect reduces the heat transfer area needed.
35. The use of high pressure steam in the first effect decreases the size and the cost of the evaporator, but steam temperature is limited by product quality considerations.
36. Last effect pressure is limited by cooling water temperature in the condenser.
37. At steady state, pressure and temperature in the intermediate effects are automatically adjusted by the process itself.
38. A liquid vapor separator is not necessary in a forced circulation evaporator.
39. A barometric leg may be used with the condenser to make easier the exit of condensate.
40. A steam jet may be used to remove noncondensables from an evaporator.
41. A vacuum pump is used to remove the vapor produced in the last effect.
42. A height of about 6 m is used in the leg of a barometric condenser.
43. Noncondensables in the steam chest decrease the condensation rate and the condensation temperature.
44. Approximately one kg of water is evaporated per kg of steam consumed in the first effect of a multiple effect evaporator.
45. Steam economy is defined as the ratio of steam consumption to the amount of water evaporated.
46. Condensate reuse may be used as a measure for energy conservation.
47. Fouling or scaling of the heat transfer surface may decrease steam economy.
48. Excess steam venting to keep noncondensables at a low level increases energy consumption.
49. A typical steam economy for a four-effect evaporator without vapor recompression is 4.3.

50. High pressure steam is used in thermal vapor recompression (TVR) to recompress part of the exit vapors.
51. The ratio of the discharge pressure to the suction pressure in the steam jet of a TVR system must be less than 1.8.
52. Because of the low temperature difference that is used with mechanical vapor recompression (MVR), the heat transfer area required is large.
53. Evaporators with TVR have a better steam economy than evaporators with MVR.
54. A four-effect evaporator with TVR in the first effect has approximately the same steam economy with a five-effect evaporator without TVR.
55. Evaporators with MVR may have a steam economy equivalent to that of a multiple-effect evaporator with 15-20 effects.

Examples

Example 16.1
Grape juice at a rate of 3 kg/s is concentrated in a single effect evaporator from 18% to 23% solids content. Calculate a) the product flow rate, b) the evaporation rate, c) the steam consumption, d) the steam economy, and e) the required heat transfer area of the evaporator if the juice enters the evaporator at 50 °C, the juice boils in the evaporator at 50 °C, saturated steam at 100 °C is used as heating medium, the condensate exits at 100 °C, the heat capacity of the juice is 3.7 kJ/kg °C and 3.6 kJ/kg °C at the inlet and the outlet of the evaporator respectively, and the overall heat transfer coefficient is 1500 W/m² °C.

Solution

Step 1
Draw the process diagram:

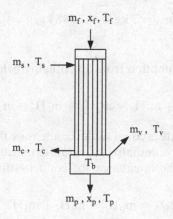

where mf, xf, Tf are the mass flow rate, solids content and temperature for the
  feed;

m_p, x_p, and T_p are the mass flow rate, solids content, and temperature for
  the product;

m_s, T_s are the mass flow rate and temperature for the steam;

m_c, T_c are the mass flow rate and temperature for the condensate;

m_v, T_v are the mass flow rate and temperature for the vapor produced;
  and

T_b is the boiling temperature.

Step 2
State your assumptions:

- Heat losses to the environment are neglected.
- The boiling point elevation of the juice is negligible.
- The evaporator operates at steady state.

Step 3
Calculate the product flow rate from a solids mass balance:

$$m_f x_f = m_p x_p$$

or

$$m_p = \frac{m_f x_f}{x_p} = \frac{(3\,\text{kg/s})\,(0.18)}{0.23} = 2.348\,\text{kg/s}$$

Step 4
Calculate the evaporation rate from a mass balance:

$$m_f = m_p + m_v$$

and

$$m_v = m_f - m_p = 3\,\text{kg/s} - 2.348\,\text{kg/s} = 0.652\,\text{kg/s}$$

Step 5
Calculate the steam consumption from an enthalpy balance:

$$m_f H_f + m_s H_s = m_p H_p + m_c H_c + m_v H_v$$

Because the steam flow rate is equal to the condensate flow rate ($m_s = m_c$) and
the difference between the enthalpy of saturated steam and the enthalpy of
condensate is equal to the latent heat of condensation ($H_s - H_c = \lambda_s$), the
above equation becomes:

$$m_s \lambda_s = m_p H_p - m_f H_f + m_v H_v$$

Solve for m_s:

$$m_s = \frac{m_p H_p - m_f H_f + m_v H_v}{\lambda_s}$$

Using $0\,^{\circ}C$ as a reference temperature, the enthalpy of the liquid streams is:

$$H = c_p T$$

The enthalpy of saturated steam, saturated vapor, and condensate is found from steam tables:

$$H_s \text{ at } 100\,^{\circ}C = 2676 \text{ kJ/kg}$$

$$H_c \text{ at } 100\,^{\circ}C = 419 \text{ kJ/kg}$$

$$\lambda_s \text{ at } 100\,^{\circ}C = 2676 - 419 = 2257 \text{ kJ/kg}$$

$$H_v \text{ at } 50\,^{\circ}C = 2592 \text{ kJ/kg}$$

Therefore:

$$m_s = \frac{m_p H_p - m_f H_f + m_v H_v}{\lambda_s} = \frac{m_p c_{pp} T_p - m_f c_{pf} T_f + m_v H_v}{\lambda_s} =$$

$$= \frac{(2.348 \text{ kg/s})(3600 \text{ J/kg}\,^{\circ}C)(50\,^{\circ}C) - (3 \text{ kg/s})(3700 \text{ J/kg}\,^{\circ}C)(50\,^{\circ}C)}{2257000 \text{ J/kg}} +$$

$$+ \frac{(0.652 \text{ kg/s})(2592000 \text{ J/kg})}{2257000 \text{ J/kg}} = 0.690 \text{ kg/s}$$

Step 6
Calculate the steam economy:

$$\text{steam economy} = \frac{m_v}{m_s} = \frac{0.652 \text{ kg/s}}{0.690 \text{ kg/s}} = 0.945 \frac{\text{kg water evaporated}}{\text{kg of steam}}$$

Step 7
Calculate the heat transfer area of the evaporator:

i) Write the heat transfer rate equation:

$$q = AU(T_s - T_b)$$

ii) Write an enthalpy balance on the steam side of the evaporator:

$$q = m_s H_s - m_c H_c = m_s \lambda_s$$

iii) Combine the last two equations:

$$m_s \lambda_s = AU(T_s - T_b)$$

iv) Solve for A, substitute values, and calculate A:

$$A = \frac{m_s \lambda_s}{U(T_s - T_b)} = \frac{(0.690 \text{ kg/s}) (2257000 \text{ J/kg})}{\left(1500 \text{ W/m}^2 \text{°C}\right) (100 - 50) \text{°C}} = 20.8 \text{ m}^2$$

Example 16.2

Tomato juice at 5.5 kg/s feed flow rate and 60 °C inlet temperature is concentrated in a double-effect forward feed evaporator using steam in the first effect at a pressure of 97.2 kPa (gauge) and cooling water in the condenser (surface condenser) entering at 30 °C and leaving at 45 °C. The heat transfer area, the overall heat transfer coefficient, the boiling point elevation (BPE), and the pressure in each effect are shown in the following table. The solids content and the heat capacity of the feed are 11% and 3900 J/kg °C respectively. Calculate the steam flow rate, the solids content at the exit of each effect, the steam economy, and the flow rate of cooling water in the condenser.

	First Effect	Second Effect
Heat transfer area, m^2	100	100
Overall heat transfer coefficient, W/m2 °C	2000	1000
Boiling point elevation, °C	0.4	0.8
Saturation pressure, kPa	90	17.9
Heat losses, kW	5	4
Heat capacity at the exit, J/kg °C	3800	3450

Solution

Step 1
Draw the process diagram:

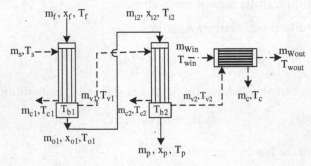

— liquid food   — — steam/vapour/condensate   — · · cooling water

Step 2
State your assumptions:

- The condensate in each effect exits at the condensation temperature.
- The operation is steady state.

Step 3
Combine the heat transfer rate equation with an enthalpy balance in the first effect, as in step 7 of Exercise 16.1, to calculate the steam consumption:

$$q = m_s H_s - m_{c1} H_{c1} = A_1 U_1 (T_s - T_{b1}) \qquad (16.1)$$

Because the condensate was assumed to exit at the condensation temperature, $H_s - H_{c1} = \lambda_s =$ the latent heat of condensation at 120 °C (97.2 kPa gauge pressure corresponds to 198.5 kPa absolute pressure or 120 °C saturation temperature in the steam chest of the first effect). Also $m_s = m_{c1}$. Therefore eqn (16.1) becomes:

$$m_s \lambda_s = A_1 U_1 (T_s - T_{b1}) \qquad (16.2)$$

with

$$H_s \text{ at } 120\,°C = 2706 \text{ kJ/kg(from Steam tables)}$$

$$H_{c1} \text{ at } 120\,°C = 504 \text{ kJ/kg(from Steam tables)}$$

$$\lambda_s \text{ at } 120\,°C = 2706 - 504 = 2202 \text{ kJ/kg}$$

The saturation temperature in the 1st effect, T_{sat1}, is found from steam tables for $P_{sat} = 90$ kPa to be equal to 96.6 °C. Therefore, the boiling temperature in the 1st effect, T_{b1}, is:

$$T_{b1} = T_{sat1} + BPE_1 = 96.6 + 0.4 = 97\,°C \qquad (16.3)$$

Solve eqn (16.2) for m_s, substitute values, and calculate m_s:

$$m_s = \frac{A_1 U_1 (T_s - T_{b1})}{\lambda_s} = \frac{(100\,m^2)\,(2000\,W/m^{2}°C)\,(120 - 97)\,°C}{2202000\,J/kg}$$

$$= 2.089 \text{ kg/s}$$

Step 4
Calculate the solids content:

1) At the exit of the first effect:

i) Write an enthalpy balance around the first effect:

$$m_f H_f + m_s H_s = m_{o1} H_{o1} + m_{c1} H_{c1} + m_{v1} H_{v1} + q_{loss\,1} \qquad (16.5)$$

ii) Analyze the terms of eqn (16.5):

a) Find H_{v1}:
The vapor exiting from the first effect is slightly superheated due to the BPE of the juice. Thus:

$$H_{v1} = H_{v1\ saturated} + c_{p\ vapour}BPE$$

Because the BPE is small, H_{v1} can be assumed equal to the enthalpy of saturated steam, $H_{v1\ saturated}$. Thus:

$$H_{v1} at 96.6\,°C = 2671 kJ/kg (from\ Steam\ tables)$$

b) Find m_{v1}:
A mass balance on the product gives:

$$m_f = m_{o1} + m_{v1}$$

or

$$m_{v1} = m_f - m_{o1} \tag{16.6}$$

c) Find the enthalpy of liquid streams:
Using $0\,°C$ as a reference temperature, the enthalpy of the liquid streams is:

$$H = c_p T \tag{16.7}$$

iii) Substitute eqns (16.6) and (16.7) in eqn (16.5):

$$m_f c_{pf} T_f + m_s H_s = m_{o1} c_{po1} T_{o1} + m_{c1} H_{c1} + (m_f - m_{o1}) H_{v1} + q_{loss\ 1} \tag{16.8}$$

iv) Solve eqn (16.8) for m_{o1}, substitute values, and calculate the mass flow rate at the exit of the first effect:

$$m_{o1} = \frac{m_f c_{pf} T_f + m_s \lambda_s - m_f H_{v1} - q_{loss\,1}}{c_{po1} T_{o1} - H_{v1}} = \frac{(5.5\,kg/s)(3900\,J/kg\,°C)\,(60\,°C)}{(3800\,J/kg°C)(97\,°C) - 2671000\,J/kg}$$

$$+ \frac{(2.089\,kg/s)\,(2202000\,J/kg) - (5.5\,kg/s)(2671000\,J/kg) - 5000\,W}{(3800\,J/kg\,°C)\,(97\,°C) - 2671000\,J/kg}$$

$$= 3.826\,kg/s$$

v) Calculate the solids content at the exit of the first effect from a solids mass balance:

$$m_f x_f = m_{o1} x_{o1} \tag{16.9}$$

Solve eqn (16.9) for x_{o1}, substitute values, and calculate x_{o1}:

$$x_{o1} = \frac{m_f x_f}{m_{o1}} = \frac{(5.5\,\text{kg/s})\,(0.11)}{3.826\,\text{kg/s}} = 0.158$$

2) At the exit of the second effect:

i) Write an enthalpy balance around the second effect:

$$m_{i2}H_{i2} + m_{v1}H_{v1} = m_p H_p + m_{c2}H_{c2} + m_{v2}H_{v2} + q_{loss\,2} \qquad (16.10)$$

ii) Analyze the terms of eqn (16.10):

$$m_{i2} = m_{o1} = 3.826\,\text{kg/s},$$

$$m_{v1} = m_f - m_{o1} = 5.5 - 3.826 = 1.674\,\text{kg/s},$$

$$m_{c2} = m_{v1} = 1.674\,\text{kg/s},$$

$$H_{i2} = H_{o1} = c_{po1}T_{o1} = (3800\,\text{J/kg}\,^\circ\text{C})\,(97\,^\circ\text{C}) = 368600\,\text{J/kg},$$

$$H_{v1} = 2671\ \text{kJ/kg(from steam tables)}$$

$$H_{c2}\ \text{at}\ 96.6\,^\circ\text{C} = 405\ \text{kJ/kg(from steam tables)}$$

The saturation temperature in the second effect, T_{sat2}, is found from steam tables for $P_{sat} = 17.9$ kPa to be equal to $57.6\,^\circ$C. Therefore, the boiling temperature in the second effect, T_{b2}, is:

$$T_{b2} = T_{sat2} + BPE_2 = 57.6 + 0.8 = 58.4\,^\circ\text{C}$$

The enthalpy, H_{v2} (neglecting the enthalpy of superheating), is:

$$H_{v2}\ \text{at}\ 57.6\,^\circ\text{C} = 2605\ \text{kJ/kg(from steam tables)}$$

Write a mass balance on the product to calculate m_{v2}:

$$m_{i2} = m_p + m_{v2}$$

Solve for m_{v2}

$$m_{v2} = m_{i2} - m_p \qquad (16.11)$$

iii) Substitute eqn (16.11) in eqn (16.10):

$$m_{i2}H_{i2} + m_{v1}H_{v1} = m_p c_{pp}T_p + m_{c2}H_{c2} + (m_{i2} - m_p)\,H_{v2} + q_{loss\,2}\quad(16.12)$$

iv) Solve eqn (16.12) for m_p and substitute values:

$$m_p = \frac{m_{i2}H_{i2} + m_{v1}H_{v1} - m_{c2}H_{c2} - m_{i2}H_{v2} - q_{loss}2}{c_{pp}T_p - H_{v2}} =$$

$$= \frac{(3.826\,\text{kg/s})\,(368600\,\text{J/kg}) + (1.674\,\text{kg/s})\,(2671000 - 405000\,\text{J/kg})}{(3450\,\text{J/kg\,°C})\,(58.4\,°\text{C}) - 2605000\,\text{J/kg}}$$

$$- \frac{(3.826\,\text{kg/s})(2605000\,\text{J/kg}) - 4000\,\text{J/s}}{(3450\,\text{J/kg\,°C})\,(58.4\,°\text{C}) - 2605000\,\text{J/kg}} = 1.983\,\text{kg/s}$$

v) Substitute values in eqn (16.11) and calculate the evaporation rate in the second effect:

$$m_{v2} = m_{i2} - m_p = 3.826 - 1.983 = 1.843\ \text{kg/s}$$

vi) Calculate the solids content at the exit of the second effect from a solids mass balance:

$$m_{fi}x_{i2} = m_p x_p$$

Solve for x_p, substitute values, and calculate x_p:

$$x_p = \frac{m_{i2}x_{i2}}{m_p} = \frac{(3.826\,\text{kg/s})\,(0.158)}{1.983\,\text{kg/s}} = 0.305$$

Step 5
Calculate the steam economy:

$$\text{steam economy} = \frac{m_{v1} + m_{v2}}{m_s} = \frac{1.674 + 1.843}{2.089}$$

$$= 1.68\,\frac{\text{kg water evaporated}}{\text{kg steam}}$$

Step 6
Calculate the cooling water flow rate in the condenser:

i) Write an enthalpy balance in the condenser:

$$m_{w\,in}H_{w\,in} + m_{v2}H_{v2} = m_{w\,out}H_{w\,out} + m_c H_c$$

or

$$m_{w\,in}c_{pw\,in}T_{w\,in} + m_{v2}H_{v2} = m_{w\,out}c_{pw\,out}T_{w\,out} + m_c H_c \quad (16.13)$$

with

$$m_{v2} = m_c = 1.843 \text{ kg/s}$$

$$H_{cl} \text{ at } 57.6°C = 241 \text{ kJ/kg(from steam tables)}$$

$$m_{w\,in} = m_{w\,out} = m_w$$

$$c_{pw\,in} \approx c_{pw\,out} \approx 4190 \text{J/kg°C}$$

ii) Solve eqn (16.13) for m_w, substitute values, and calculate the cooling water flow rate:

$$m_w = \frac{m_{v2}(H_{v2} - H_c)}{c_{pw}(T_{w\,out} - T_{w\,in})} = \frac{(1.843 \text{ kg/s})\,(2605000 - 241000)\text{J/kg}}{(4190 \text{ J/kg°C})(45 - 30)\,°C} = 69.3 \text{ kg/s}$$

Exercises

Exercise 16.1

Calculate the steam consumption in a single effect evaporator with 25 m² heat transfer area, which is being used to concentrate a fruit juice. The juice enters the evaporator at 70 °C, the saturation pressure in the evaporator is 31.19 kPa, saturated steam at 100 °C is used as the heating medium, the condensate exits at 95 °C, and the overall heat transfer coefficient is 1500 W/m2 °C.

Step 1
Draw the process diagram:

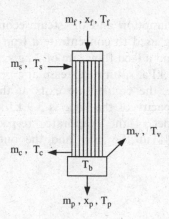

Step 2
State your assumptions:

- The boiling point elevation is negligible.
- The heat losses to the environment are negligible.
- The system operates at steady state.

Step 3
Write the heat transfer rate equation:

$$\text{...} \tag{16.14}$$

Step 4
Write an enthalpy balance on the steam side:

$$\text{...} \tag{16.15}$$

Step 5
Combine eqns (16.14) and (16.15) and calculate the steam consumption:

$$m_s = \text{..}$$

Exercise 16.2

Calculate the steam consumption and the steam economy in a single effect evaporator which is being used to concentrate a fruit juice. The juice enters the evaporator at 25 °C and a feed flow rate of 2 kg/s, the saturation pressure in the evaporator is 31.19 kPa, saturated steam at a pressure of 143.27 kPa is used as heating medium, the condensate exits at the steam condensation temperature, the heat capacity of the juice is 3.9 kJ/kg °C and 3.7 kJ/kg °C at the inlet and the outlet of the evaporator respectively, and the solids content is 10% and 20% at the inlet and the outlet of the evaporator respectively.

Solution

Step 1
Draw the process diagram:

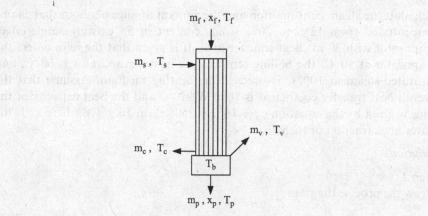

Step 2
State your assumptions:

....

Step 3
Write a solids mass balance to calculate the outlet flow rate:

$$m_p = \text{...}$$

Step 4
Write an enthalpy balance around the effect:

$$m_f H_f + m_s H_s = \text{..}$$ (16.16)

Find:

H_v at 31.19 kPa = ...kJ/kg

H_s at 143.27 kPa = ...kJ/kg

H_c at 143.27 kPa = ...kJ/kg

Solve eqn (16.16) for m_s, substitute values, and calculate m_s:

$$m_s = \text{...}$$

Step 5
Calculate the steam economy:

steam economy = ...

Exercise 16.3

Calculate the steam consumption and the amount of juice per hour that can be concentrated from 12% to 20% solids content in an existing single effect evaporator with 30 m² heat transfer area. It is given that the juice enters the evaporator at 50 °C, the boiling temperature in the evaporator is 60 °C, and saturated steam at 100 °C is used as the heating medium. Assume that the overall heat transfer coefficient is 1000 W/m² °C and the heat capacity of the juice is given by the equation $c_p = 1672 + 2508 \, x_w$ (in J/kg °C), where x_w is the water mass fraction of the juice.

Solution

Step 1
Draw the process diagram:

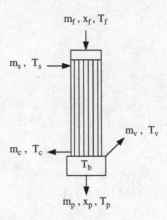

Step 2
State your assumptions:

- The condensate exits at the steam condensation temperature.

- ....

Step 3
Calculate the steam consumption from the heat transfer rate equation and an enthalpy balance on the steam side:

$$.. \tag{16.17}$$

Find:

H_s at 100 °C = ...kJ/kg

H_c at 100 °C = ...kJ/kg

Solve eqn (16.17) for m_s, substitute values, and calculate the steam consumption:

$m_s =$ ....

Step 4
Calculate the feed flow rate, m_f:

    i) Express m_p as a function of m_f from a solids mass balance:

$$m_p = \text{..} \qquad (16.18)$$

    ii) Write an enthalpy balance around the effect:

$$m_f c_{pf} T_f + \text{...} \qquad (16.19)$$

    iii) Substitute m_p from eqn (16.18) in eqn (16.19):

$$m_f c_{pf} T_f + \text{...} \qquad (16.20)$$

    iv) Find:

$$H_v \text{ at } 60\,^{\circ}C = \text{................................} kJ/kg$$

$$c_{pf} = \text{...} J/kg\,^{\circ}C$$

$$c_{pp} = \text{...} J/kg\,^{\circ}C$$

    v) Solve eqn (16.20) for m_f, substitute values, and calculate m_f:

$m_f =$ ....

Exercise 16.4

A fruit juice is concentrated from 10% to 30% solids content in a double-effect forward-feed evaporator at a flow rate of 10000 kg/h. The temperature of the juice at the inlet is 70 °C and the steam temperature is 110 °C. The cooling water temperature at the inlet of the condenser (surface condenser) is 25 °C and at the outlet 42 °C. The heat capacity of the juice at the inlet of the evaporator is 3.9 kJ/kg °C. Calculate the evaporation rate in each effect, the steam economy, and the cooling water flow rate in the condenser. The heat capacity at the exit of each effect and the boiling temperature are given in the following table:

	First Effect	Second Effect	Condenser
Heat capacity at the exit, J/kg °C	3800	3400	4190
Boiling temperature, °C	87	55	

Solution

Step 1
Draw the process diagram:

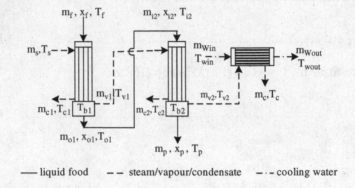

——— liquid food   – – steam/vapour/condensate   – ·– cooling water

Step 2
State your assumptions:

....

Step 3
Calculate the evaporation rate.

i) Write a solids mass balance to calculate the outlet flow rate from the first effect:

$$m_p = ... \quad (16.21)$$

ii) Write an enthalpy balance around the second effect to calculate the flow rate at the inlet of the second effect m_{i2}, taking into account that $m_{i2} = m_{o1}$:

$$m_{i2}H_{i2} + m_{v1}H_{v1} = ... \quad (16.22)$$

Find:

$$H_v \text{ at } 87\,^\circ C = \text{................................}kJ/kg$$

$$H_v \text{ at } 55\,^\circ C = \text{................................}kJ/kg$$

$$H_c \text{ at } 87\,^\circ C = \text{................................}kJ/kg$$

$$H_c \text{ at } 55\,^\circ C = \text{................................}kJ/kg$$

iii) Write a mass balance on the liquid food side in the first effect to calculate the evaporation rate in the first effect:

$$m_{v1} = \text{...} \qquad (16.23)$$

iv) Write a mass balance on the liquid food side in the second effect to calculate the evaporation rate in the second effect:

$$m_{v2} = \text{...} \qquad (16.24)$$

v) Combine eqns (16.22), (16.23) and (16.24) and solve for mo_1

$$mo_1 = \text{...}$$

Step 4
Calculate the steam economy:

i) Write an enthalpy balance around the first effect to calculate the steam flow rate in the first effect, m_s:

$$m_f H_f + m_s H_s = \text{...} \qquad (16.25)$$

Find:

$$H_s \text{ at } 100\,^\circ C = \text{................................}kJ/kg$$

$$H_c \text{ at } \text{..........} = \text{................................}kJ/kg$$

Solve eqn (16.25) for m_s and substitute values

$$m_s = \text{...}(16.26)$$

ii) Calculate the steam economy from eqns (16.23), (16.24), and (16.26):

$$\text{steam economy} = \text{...}$$

Step 5
Write an enthalpy balance around the condenser to calculate the mass flow rate of the cooling water:

$$m_{win}H_{win} + \dots\dots\dots\dots\dots\dots\dots\dots\dots\dots\dots\dots\dots\dots\dots\dots\dots\dots\dots \quad (16.27)$$

Solve eqn (16.27) for m_{win} and substitute values:

$$m_{win} = \dots\dots\dots\dots\dots\dots\dots\dots\dots\dots\dots\dots\dots\dots\dots\dots\dots\dots\dots$$

Exercise 16.5

Solve the previous problem for a double-effect backward-feed evaporator, assuming that the heat capacity is 3800 J/kg °C at the exit of the second effect and 3400 J/kg °C at the exit of the first effect.

Solution

Step 1
Draw the process diagram:

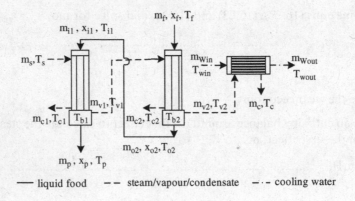

— liquid food   — — steam/vapour/condensate   — ·· cooling water

Step 2
State your assumptions:

• ....

• ....

Step 3
Calculate m_p with a mass balance.

Step 4
Calculate m_{v1} and m_{v2}, solving simultaneously an enthalpy balance and mass balances.

Step 5
Calculate the steam consumption with an enthalpy balance.

Step 6
Calculate the steam consumption.

Step 7
Calculate the cooling water flow rate with an enthalpy balance.

Exercise 16.6

Calculate the required heat transfer area for each effect in Exercise 16.4 if the overall heat transfer coefficient is 1500 $W/m^2\,°C$ and 1000 $W/m^2\,°C$ in the first and second effects respectively.

Solution

Step 1
Write the heat transfer rate equation for the first effect and solve for A_1.

Step 2
Write the heat transfer rate equation for the second effect and solve for A_2.

Exercise 16.7

Solve Exercise 16.4 using the spreadsheet program *Evaporator.xls*.

Solution
In the sheet "2-Effect Forward", insert in the yellow cells the data given in Exercise 16.4 and follow the instructions. For the heat transfer area and the overall heat transfer coefficient of the evaporator and the condenser use your own values. Indicative values are $A_1 = A_2 = 80m^2$ in cells H19 and J19, $A_c = 50m^2$ in cell M19, $U_1 = 1200 W/m^2\,°C$ in cell H20, $U_2 = 800 W/m^2\,°C$ in cell J20, $U_c = 2500 W/m^2\,°C$ in cell M20.
Rerun the program using a smaller heat transfer area in the effects, e.g., 70 m^2. Observe how the amount of cooling water, the temperature of cooling water at the exit of the condenser, and the boiling temperature in each effect are affected. Explain the results.
Rerun the program using 120 °C as steam temperature. Explain the results.
Rerun the program using 100 m^2 as condenser surface area. Explain the results.

Exercise 16.8

Milk is concentrated in a four-effect forward-feed evaporator from an initial solids content of 9% to a final solids content of 48%. The feed flow rate is 20000 kg/h, the inlet temperature of the milk in the first effect is 60 °C, the steam temperature is 75 °C, and the cooling water temperature is 25 °C. The evaporator is equipped with a contact condenser with 150 m^2 heat transfer area for the condensation of the vapors exiting the fourth effect. The overall heat transfer coefficient in the condenser is 2500 W/m2 °C. Find the required heat transfer area of each effect and the cooling water flow rate, so that the boiling temperature in the first effect is 70 ± 1 °C and in the last effect is 40 ± 1 °C. Calculate the steam economy. Assume equal heat transfer area in the effects. Assume that the overall heat transfer coefficient changes as a function of temperature and concentration according to: $U = 400\exp(0.03T)\exp(-2.5x_s)$, where U is W/m^2 °C, T is temperature in °C, and x_s is the solids mass fraction.

Solution

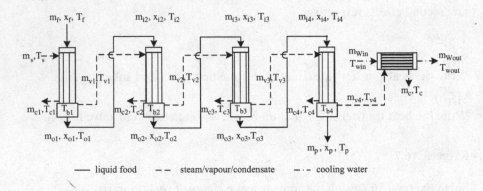

Solve the problem using the spreadsheet program *Evaporator.xls*. In the sheet "4-Effect Forward", insert the data in the yellow cells and follow the instructions. Insert a trial value for the heat transfer area in cells H18, J18, L18 and N18 and iterate until a "FINISH" signal is shown in the control panel. Check if the boiling temperature in cells H23 and N23 is 70 ± 1°C and 40 ± 1 °C. If not, use another trial value for the heat transfer area and iterate. Repeat this procedure until the boiling temperature is equal to the given values.

Observe how the overall temperature difference changes as the heat transfer area changes. Explain this.

Observe how the saturation pressure changes from effect to effect. Explain this.

Observe how the boiling temperature in the effects changes as the cooling water flow rate changes. Explain this.

Explain why the evaporation rate increases from the first effect to the fourth effect.

Increase the heat transfer area of the condenser to 200 m^2 and rerun the program using the heat transfer area found above for each effect. Observe how the cooling water flow rate changes. Explain the results.

Increase the steam temperature to 80 °C. Rerun the program. Observe how the boiling temperature and the cooling water flow rate change. Explain the results.

Exercise 16.9

Solve Exercise 16.8 for the case of a backward-feed four-effect evaporator and for the case of a forward-feed two-effect evaporator using the corresponding spreadsheets in the program evaporator.xls. Use the area found for the forward-feed evaporator as the heat transfer area for the effects. Compare the results. Explain the differences.

Decrease the feed temperature to 30 °C. Rerun the program. Compare the results. Explain the difference in the steam economy between the forward-feed four-effect and the backward-feed four-effect evaporators.

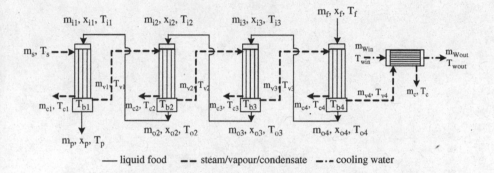

Four-effect, backward-feed evaporator

Chapter 17
Psychrometrics

Review Questions

Which of the following statements are true and which are false?

1. The moisture content of air is also called the humidity ratio or absolute humidity.
2. The humidity ratio of a given air-water vapor mixture increases as the temperature of the mixture increases.
3. The humidity ratio of an air-water vapor mixture is independent of the partial pressure of water vapor in the mixture.
4. The relative humidity of an air-water vapor mixture is also called percentage saturation or percentage absolute humidity.
5. The relative humidity of an air-water vapor mixture is independent of the temperature of the mixture.
6. The wet bulb temperature of an air-water vapor mixture is higher than the dry bulb temperature.
7. The wet bulb temperature of an air-water vapor mixture is approximately equal to the adiabatic saturation temperature.
8. During an adiabatic saturation process, the sensible heat of an air-water vapor mixture decreases but its latent heat increases.
9. The wet bulb temperature of an air-water vapor mixture is equal to the dry bulb temperature when the relative humidity is equal to 100%.
10. The dry bulb temperature of an air-water vapor mixture is equal to the dew point temperature when the relative humidity is equal to 100%.
11. When a saturated air-water vapor mixture is cooled, some water vapor will condense.
12. The humid heat of an air-water vapor mixture contains the sensible heat of dry air and the latent heat of the water vapor.
13. The humid volume of an air-water vapor mixture is equal to the inverse of the density of the mixture.
14. The properties of an air-water vapor mixture can be determined from a psychrometric chart if two independent property values are known.
15. Every psychrometric chart is drawn for a specific barometric pressure.

16. The wet bulb lines coincide with constant enthalpy lines.
17. The psychrometric ratio is a function of the heat and the mass transfer coefficients.
18. The psychrometric ratio of an air-water vapor mixture is equal to the humid heat of that mixture.
19. The relative humidity of air can be determined if its dry bulb and wet bulb temperatures are known.
20. When air is cooled below its dew point temperature, both sensible heat and latent heat are given off.

Examples

Example 17.1

Find the properties of air with 40 °C dry bulb temperature and 28.5 g water/kg dry air moisture content at 1 atm.

Solution

Step 1

i) Locate the point on the psychrometric chart with 40 °C dry bulb temperature and 28.5 g moisture/kg dry air moisture content.
ii) Draw a horizontal line that passes from the point with 28.5 g water/kg dry air moisture content on the y-axis.
iii) Draw a vertical line on the x-axis at the point of 40 °C dry bulb temperature. The point the two lines intersect (point A) represents the point on the psychrometric chart with 40 °C dry bulb temperature and 28.5 g water/kg dry air moisture content.

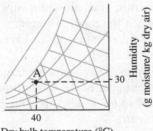

Step 2

i) Find the relative humidity curve that passes through point A.
ii) Read the relative humidity on the relative humidity curve: Read RH = 60%.

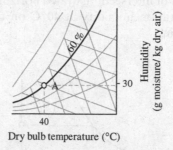

Dry bulb temperature (°C)

Step 3

i) Draw a line parallel to the wet bulb line that passes through point A.
ii) Read the wet bulb temperature at the point where this line crosses the saturation curve (100% relative humidity curve): Read T_w = 32.5 °C. Alternatively, draw a vertical line down from that point to the dry bulb temperature axis and also read 32.5 °C on the x-axis.

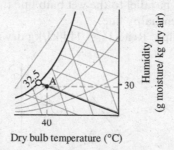

Dry bulb temperature (°C)

Step 4

i) Draw a line parallel to the x-axis that passes through the wet bulb temperature of 32.5 °C found above.
ii) Read the saturation humidity at the wet bulb temperature on the y-axis: Read W_w = 0.032 kg water/kg dry air.

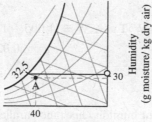

Dry bulb temperature (°C)

Step 5

i) Draw a vertical line from the x-axis that passes through point A to cross the 100% relative humidity curve.
ii) From the point of intersection, draw a horizontal line to cross the y-axis.
iii) Read the saturation humidity at 40 °C on the y-axis. Read: 0.049 kg water/kg dry air.

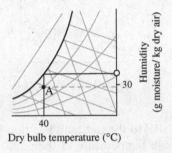

Step 6

i) Extend the line parallel to the wet bulb line that passes through point A to cross the enthalpy line.
ii) Read the enthalpy: Read H = 114 kJ/kg dry air.

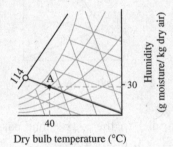

Alternatevely, calculate enthalpy from:

$$H = (1.005 + 1.88W)T + \lambda_o W = (1.005 + 1.88*0.0285)40 + 2501*$$
$$0.0285 = 113.6 \, kJ/kg \, d.a.$$

Step 7

i) Extend the constant humidity line (line parallel to the x-axis) that passes through point A to cross the saturation curve (100% relative humidity curve).

ii) Read the dew-point temperature on the saturation line: Read $T_{dp}= 30.5°C$.

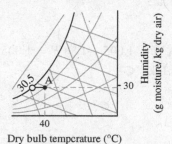

Step 8

i) Draw a line parallel to the humid volume line that passes through point A.

ii) Read the humid volume: read $v_H= 0.93m3/kg$ dry air.

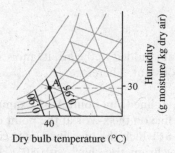

iii) Alternatively, calculate the humid volume from:

$$v_H = (0.00283 + 0.00456W)T = (0.00283 + 0.00456 \times 0.0285)$$
$$(273 + 40) = 0.926\,m^3/kg\,dry\,air$$

Step 9

Calculate the percentage saturation or percentage absolute humidity:

$$PS = 100\frac{W}{W_s} = \frac{0.0285}{0.049} = 58.2\%$$

Step 10

Calculate the partial pressure of water vapor in the air:
Since:

$$W = \frac{18.02}{28.96}\frac{p}{P - p} = 0.622\frac{p}{P - p}$$

$$p = \frac{WP}{0.622 + W} = \frac{(0.0285)(101325)}{0.622 + 0.0285} = 4439.3\,Pa$$

Step 11
Calculate the humid heat:

$$c_s = c_A + c_V W = 1.005 + 1.88W = 1.005 + 1.88 \times 0.0285$$
$$= 1.059 \, kJ/kg \, °C$$

Example 17.2

The air of Example 17.1 is heated to 80 °C. Find the relative moisture content, wet bulb temperature, dew-point temperature, humid volume, and enthalpy of the air.

Solution

Step 1
Since only sensible heat is added to the air, its moisture content and dew-point temperature remain constant. Therefore:

i) Draw a horizontal line from point A of Example 17.1.
ii) Draw a vertical line on the x-axis at the point of 80 °C dry bulb temperature. The point at which the two lines intersect (point B) represents the point on the psychrometric chart with 80 °C dry bulb temperature and 28.5 g water/kg dry air moisture content.

Step 2
Find the properties of air at point B following the same procedure as in Example 17.1 for point A.

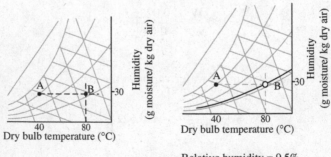

Relative humidity = 9.5%

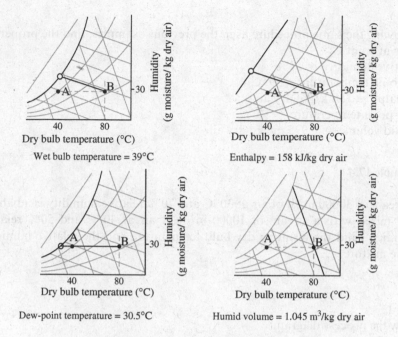

Wet bulb temperature = 39°C

Enthalpy = 158 kJ/kg dry air

Dew-point temperature = 30.5°C

Humid volume = 1.045 m³/kg dry air

Example 17.3

The heated air of Example 17.2 passes through a dryer, picking up moisture adiabatically there, and leaves the dryer at 50 °C. Determine the properties of the air at the exit of the dryer.

Solution

Step 1
The air flows through the dryer following an adiabatic saturation line. The point of intersection (point C) of the adiabatic saturation line passing through point B with the vertical line on the x-axis passing through 50 °C dry bulb temperature represents the air at the exit of the dryer.

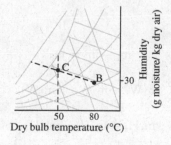

Step 2
Following the same procedure as in the previous examples, find the properties
of air at point C:
Relative humidity = 51%
Wet bulb temperature = 39 °C
Enthalpy = 158 kJ/kg dry air
Dew-point temperature = 37 °C
Humid volume = 0.975 m^3/kg dry air

Example 17.4

A stream of 30000 m^3/h of air at 30 °C and 70% relative humidity is adiabatically mixed with a stream of 10000 m^3/h of air at 40 °C and 50% relative
humidity. Find the resulting dry bulb temperature and the relative humidity
of the mixture.

Solution

Step 1
Draw the process diagram:

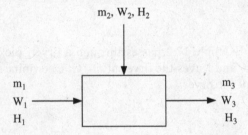

Step 2
State your assumptions: The total pressure of the air is 1 atm.

Step 3

   i) Write a total mass balance, a moisture balance, and an enthalpy balance
     for the air streams:

$$m_1 + m_2 = m_3 \qquad (17.1)$$

$$m_1 W_1 + m_2 W_2 = m_3 W_3 \qquad (17.2)$$

$$m_1 H_1 + m_2 H_2 = m_3 H_3 \qquad (17.3)$$

ii) Combine eqns (17.1), (17.2), and (17.3):

$$\frac{W_2 - W_3}{W_3 - W_1} = \frac{m_1}{m_2} \qquad (17.4)$$

and

$$\frac{H_2 - H_3}{H_3 - H_1} = \frac{m_1}{m_2} \qquad (17.5)$$

iii) Equate eqns (17.4) and (17.5):

$$\frac{m_1}{m_2} = \frac{W_2 - W_3}{W_3 - W_1} = \frac{H_2 - H_3}{H_3 - H_1} \qquad (17.6)$$

The points 1, 2, and 3 representing the air streams 1, 2, and 3 can be located on a humidity-enthalpy chart:

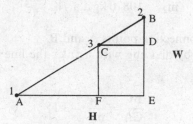

Because of eqn (17.6):

$$\frac{m_1}{m_2} = \frac{BD}{CF} = \frac{DC}{FA} \qquad (17.7)$$

Therefore, the triangles BDC and CFA are geometrically similar and the line segments BC and CA must have the same slope. This shows that the points A, B, and C must lie on a straight line. Point C can be located on the line segment AB so that:

$$\frac{BC}{CA} = \frac{m_1}{m_2}$$

Using eqn (17.6) and the properties of similar triangles, other useful relationships result:

$$\frac{m_1}{m_3} = \frac{W_2 - W_3}{W_2 - W_1} = \frac{H_2 - H_3}{H_2 - H_1} = \frac{BD}{BE} = \frac{CD}{AE} = \frac{BC}{AB} \qquad (17.8)$$

$$\frac{m_2}{m_3} = \frac{W_3 - W_1}{W_2 - W_1} = \frac{H_3 - H_1}{H_2 - H_1} = \frac{CF}{BE} = \frac{AF}{AE} = \frac{AC}{AB} \qquad (17.9)$$

Step 4

To solve the problem:

i) Locate points A and B on the psychrometric chart.
ii) Read the humid volume of each stream: $v_{H1} = 0.88$ m^3/kg d.a. and v_{H2} $= 0.92$ m^3/kg d.a.
iii) Calculate the mass flow rate:

$$m_1 = \frac{Q_1}{v_{H1}} = \frac{30000 \text{ m}^3/\text{h}}{0.88 \text{ m}^3/\text{kg d.a.}} = 34091 \text{ kg d.a./h}$$

$$m_2 = \frac{Q_2}{v_{H2}} = \frac{10000 \text{ m}^3/\text{h}}{0.92 \text{ m}^3/\text{kg d.a.}} = 10870 \text{ kg d.a./h}$$

and

$$\frac{m_1}{m_2} = \frac{34091 \text{ kg d.a./h}}{10870 \text{ kg d.a./h}} = 3.1$$

iv) Draw the line connecting points A and B.
v) Locate point C by dividing with a ruler the line segment AB into two parts such that:

$$\frac{BC}{CA} = \frac{m_1}{m_2} = 3.1$$

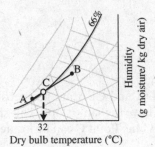

vi) Read the properties of air at point C as:
Dry bulb temperature = 32 °C,
Relative humidity = 66%.

Alternative solution

Step 1

Divide eqn (17.3) by m_3:

$$\frac{m_1}{m_3}H_1 + \frac{m_2}{m_3}H_2 = H_3 \tag{17.10}$$

Step 2
Read the enthalpy of points A and B from the psychrometric chart.

$$H_1 = 78 \text{ kJ/kg d.a. and } H_2 = 101 \text{ kJ/kg d.a.}$$

Step 3
i) Find m_3 using eqn (17.1):

$$m_3 = 34091 \text{ kh/h} + 10870 \text{ kg/h} = 44961 \text{ kg/h}$$

ii) Calculate H_3 using eqn (17.10):

$$\frac{34091 \text{ kg d.a./h}}{44961 \text{ kg d.a./h}}(78 \text{ kJ/kg d.a.}) + \frac{10870 \text{ kg d.a.}}{44961 \text{ kg d.a.}}$$
$$\times (101 \text{ kJ/kg d.a.}) = 83.5 \text{ kJ/kg d.a.}$$

Step 4
On the psychrometric chart, locate point C, which is the point of intersection of line segment AB and the isoenthalpic line with H = 83.5kJ/kg d.a.

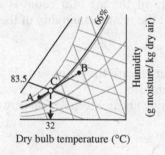

Dry bulb temperature (°C)

Step 5
Read the properties of air at point C as:
    Dry bulb temperature = 32 °C
    Relative humidity = 66%

Comment: Eqn (17.2) instead of eqn (17.3) could be used and work with absolute humidity instead of enthalpy.

Exercises

Exercise 17.1

For air at 80 °C dry bulb temperature and 20 °C dew-point temperature, determine its humidity, relative humidity, wet bulb temperature, enthalpy, humid volume and humid heat.

Solution

On the psychrometric chart locate the point with 80 °C dry bulb temperature
and 20 °C dew-point temperature.

Read the humidity on the humidity axis:.. ...
Read the relative humidity on the relative humidity curve: ....
Read the wet bulb temperature on the saturation curve: ....
Read the enthalpy on the enthalpy line: ....
Also calculate the enthalpy from the equation given in step 6 of Example 17.1:
....
Read the humid volume, interpolating between the humid volume lines: .. ..
Also calculate the humid volume from the equation given in step 8 of
Example 17.1:
....
Calculate the humid heat from the equation given in step 11 of Example 17.1:
....

Exercise 17.2

The dry bulb temperature of the air in a room is 25 °C and the wet bulb
temperature is 20 °C. Find the relative humidity of the air.

Solution

On the psychrometric chart locate the point with 25 °C dry bulb temperature
and 20 °C wet bulb temperature.
Read the relative humidity by linear interpolation between the relative humidity
curves:...................................

Exercise 17.3

Use the spreadsheet program *Moist air properties.xls* to find the properties of
the air of Example 17.1, Example 17.2, and Example 17.3.

Solution

For Example 17.1:
Insert the dry bulb temperature (40 °C) and the humidity (0.0285 kg water/
kg d.a.) of the air in cells C8 and C10 respectively. Run the program,
following the instructions.

For Example 17.2:
Insert the dry bulb temperature (80 °C) and the humidity (0.0285 kg water/
kg d.a.) of the air in cells C8 and C10 respectively. Run the program,
following the instructions.

For Example 17.3:
Insert the dry bulb temperature (50 °C) and the wet bulb temperature (39 °C) of the air in cells C8 and C9 respectively. Run the program, following the instructions.

Exercise 17.4

The dry bulb temperature of the air in a room is 18 °C and the wet bulb temperature is 15 °C. Find a) the relative humidity of the air, and b) the new value of the relative humidity, if the temperature of the room increases to 23 °C. Did the dew-point temperature change?

Solution

Use the spreadsheet program *Moist air properties.xls* to find the properties of the air.

a) Insert the given values and read: Relative humidity = ...
b) Relative humidity = ...

Exercise 17.5

Air is being drawn over the cold surface of the cooling coil of an air conditioning unit. If 10 m³/min enters the air conditioner, calculate the mass flow rate of the condensed moisture, given that the air enters at 20 °C and 70% relative humidity and exits at 12 °C dry bulb temperature and 90% relative humidity. Calculate the refrigeration capacity of the coil in W and Btu/h.

Solution

Step 1
Draw the process diagram:

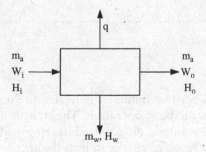

Step 2
State your assumptions:

- The system operates at steady state.
- The condensed water exits at the exit air temperature.

Step 3
Use the psychrometric chart or the spreadsheet program *Moist air properties.xls* to find the humidity, the enthalpy and the humid volume of the air at the inlet as well as the humidity and the enthalpy of the air at the outlet.

Step 4
Write a mass balance on the moisture:

$$m_a W_i = \text{...} + \text{..}$$

Step 5
Calculate m_a:

....

Step 6
Calculate m_w:

....

Step 7
Write an enthalpy balance over the cooler:

$$m_a H_i = \text{....................................} + \text{.......................} + \text{...............................}$$

Step 8
Substitute values and solve for q:

$$q = \text{...}$$

Exercise 17.6

Air at 20 °C and 60% relative humidity is heated to 90 °C and then enters a continuous dryer in an adiabatic operation. The air exits at 40 °C. Calculate: a) the inlet volumetric flow rate of air required to remove 20 kg of water/h from the product and b) the required energy to heat the air.

Solution

Step 1
Draw the process diagram:

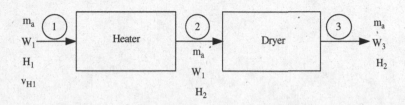

Step 2
Find the properties of the air at the inlet using a psychrometric chart or the spreadsheet program *Moist air properties.xls*.
Air at 20 °C dry bulb temperature and 60% relative humidity:

Humidity W_1 = ...
Enthalpy H_1 = ...
Humid volume v_{H1} = ...

Air at 90 °C dry bulb temperature and humidity W_1:

Enthalpy H_2 = ...
Air at 40 °C dry bulb temperature and enthalpy H_2:

Humidity W_3 = ...

Step 3
Write a mass balance on the moisture between points 2 and 3:

$$m_a(W_3 - W_1) = ...$$

Substitute values and calculate m_a:

$$m_a = ...$$

Use the humid volume to calculate the volumetric flow rate:
....

Step 4
Write an enthalpy balance on the air between points 1 and 2 and calculate the required energy:
....

Chapter 18
Drying

Review Questions

Which of the following statements are true and which are false?

1. If the drying rate is high, case hardening may result.
2. Case hardening is reduced if high humidity air is used in the dryer.
3. The free moisture content in a product is independent of the drying air temperature.
4. Only a fraction of the free moisture content can be removed with drying.
5. Unbound moisture content exerts a vapor pressure equal to the vapor pressure of pure water at the same temperature.
6. Unbound moisture may move to the surface through capillaries.
7. If moisture moves to the surface through capillaries, the drying time is proportional to the square of the thickness of the solid.
8. The critical moisture content in a product is independent of the initial drying rate.
9. At a moisture content less than the critical moisture content, the surface of the product still remains fully wet.
10. The equilibrium moisture content of a product in a dryer is independent of the air temperature in the dryer.
11. The equilibrium moisture content of a product in a dryer can be predicted if the desorption isotherm of the product at the same temperture is known.
12. The equilibrium moisture content of a product in a dryer can be predicted if the adsorption isotherm of the product at the same temperature is known.
13. The GAB model successfully fits experimental data of equilibrium moisture content for water activity values up to 0.9.
14. As the moisture in a product approaches the equilibrium moisture content, the rate of drying tends toward zero.
15. The drying rate in a constant rate period is controlled by the internal resistance to heat and mass transfer.
16. The rate of heat transfer inside a product during drying is limited by the heat transfer coefficient at the surface.

17. Air velocity significantly affects the drying rate at the constant rate period.
18. The driving force for mass transfer in the dried layer inside a product during the falling rate period is the water vapor pressure gradient.
19. The heat and mass transfer coefficients in a fluidized bed dryer are higher than in a tunnel dryer.
20. Because the air temperature in spray drying is high, the product quality is inferior.
21. The residence time of the product in a spray drier is in the range of a few seconds.
22. The droplet size in a spray dryer depends only on the type of atomizer used and is independent of the physical properties of the product.
23. Shrinkage of a freeze dried product may be significant.
24. Rehydration of freeze dried products is difficult.
25. The rate of drying in freeze drying is mainly controlled by the external resistance to heat and mass transfer.
26. The rate of drying in spray drying is mainly controlled by the external resistance.
27. The product surface temperature during the constant rate period is always equal to the wet bulb temperature of drying air.
28. The drying cost in a spray dryer is lower than in a freeze dryer.
29. Products dried by explosion puff drying have good rehydration properties.
30. The drying time in the falling rate period will be proportional to the square of the thickness of the product, even if the diffusion coefficient varies with the moisture content.
31. Cocurrent flow in a continuous dryer gives faster drying than counter-current flow.
32. Cocurrent flow results in less heat damage to the product than counter-current flow for heat sensitive products.
33. Drying time is proportional to the thickness of a product if external mass transfer controls.
34. Recirculation of a fraction of the air in a dryer increases thermal efficiency.
35. Recirculation of a fraction of the air in a dryer decreases drying time.
36. Superheated steam is not a suitable drying medium for drying foodstuffs.
37. The use of superheated steam as a drying medium increases energy consumption.
38. Vegetables may be dried in belt dryers.
39. Pretreatment of fruits and vegetables with sulphur dioxide prevents none-nzymatic browning during drying.
40. Products such as milk, coffee, and eggs are dried in spray dryers.
41. Pneumatic dryers are used for drying powders.
42. Explosion puff dryers may be used as secondary dryers in the finish drying of pieces of fruits and vegetables.
43. Instant potato powder may be produced in a drum dryer.
44. Freeze drying may be used to dry premium coffee.
45. Peas may be dried in a fluidized bed dryer.

46. Fluidized bed dryers are suitable for drying food particles up to 30 mm in size.
47. Milk dried in drum dryers is useful in the prepared-food industry because it has high water-binding capacity but it has cooked flavor.
48. Air temperatures of 120–250 °C at the inlet and 70–95 °C at the outlet are used in spray dryers.
49. Roughly 2 to 3 kg of steam are needed in a spray dryer to evaporate 1 kg of moisture from a product.
50. Spray drying is used to dry egg products after treatment with bakers yeast or glucose oxidase-catalase enzyme to remove glucose before drying.

Examples

Example 18.1

A drying test was carried out in a laboratory dryer to determine the drying rate curve and the critical moisture content of a vegetable. The sample was cut into small cubes and placed in a tray with 0.4cm × 0.4cm surface area. The tray was placed in a dryer for 17 hours and dried at constant drying conditions with an air stream that was flowing parallel to the surface of the tray. From time to time, the tray was weighed, while in a separate experiment the initial moisture content of the sample was determined. If the initial moisture content of the sample was 89% (wet basis) and the weight of the product in the tray varied with time as given below, a) plot the drying curve and the drying rate curve, and b) estimate the rate of drying at the constant-rate period, the critical moisture content, the equilibrium moisture content, and the free moisture content under the given drying conditions.

Time (min)	Net weight, m_p (kg)
0	5.000
30	3.860
60	2.700
90	2.250
120	1.900
180	1.332
240	1.164
300	1.020
360	0.912
480	0.816
540	0.780
600	0.720
660	0.696
720	0.684
780	0.655

(continued)

Time (min)	Net weight, m_p (kg)
840	0.640
900	0.630
960	0.622
1020	0.620
1080	0.620

Solution

Step 1
Calculate the kg of dry solids that are contained in the product:

$$m_s = m_p - m_{moisture} = 5 - 5 \cdot 0.89 = 0.55 \, kg \, d.s$$

Step 2
Calculate the moisture content on a dry basis for each weight measurement from:

$$X = \frac{m_p - m_s}{m_s}$$

Step 3
Plot X vs t :

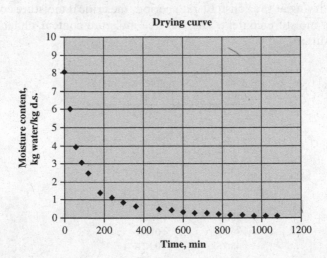

Step 4
Find the slope of the curve of X vs t at various time values, or calculate $\frac{\Delta X}{\Delta t}$ vs t or $\frac{\Delta X}{\Delta t}$ vs X from the experimental data.

Step 5

Plot $\frac{\Delta X}{\Delta t}$ vs X and read from the plot the critical moisture content X_c, the drying rate R_c at the constant-rate period, and the equilibrium moisture content X_e on the x-axis (moisture content at drying rate = 0) or from the given data.

Read:

$$R_c = 0.24 \, kg \, water/min \, m^2$$

$$X_c = 4.95 \, kg \, water/kg \, d.s.$$

$$X_e = 0.127 \, kg \, water/kg \, d.s.$$

Step 6

Calculate the free moisture content at the beginning of drying (moisture that can be removed from the solid under the given drying air conditions of temperature and humidity):

$$X_{free} = X_o - X_e = 8.09 - 0.127 = 7.964 \, kg \, water/kg \, d.s.$$

It is suggested that steps 1 to 5 be carried out on a spreadsheet. See the spreadsheet *Drying 1.xls*.

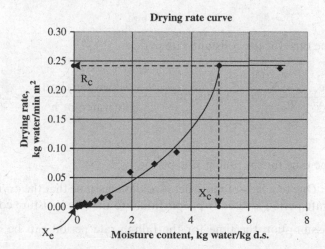

Comments:

1) The linear part of the falling rate period is missing in this product,
2) The equilibrium moisture content X_e could also be determined from the desorption isotherm of the product (if available) at the same temperature as the temperature of the air in the dryer.

Example 18.2

Calculate the time necessary to dry a product from 90% to 25% moisture (wet basis) in an industrial dryer where 2 kg dry solid/m² surface area exposed to the air is loaded. It is given that the critical moisture content is 5 kg water/ kg dry solid, the equilibrium moisture content is 0.033 kg water/ kg dry solid, and the drying rate at the critical moisture content is 3 kg water/m²h under the specified drying conditions.

Solution

Step 1
Express the moisture content on a dry basis:

i) Initial moisture content:

$$X_o = \frac{Y_o}{1 - Y_o} = \frac{0.9}{1 - 0.9} = 9 \text{ kg water/kg d.s.}$$

ii) Final moisture content:

$$X_f = \frac{Y_f}{1 - Y_f} = \frac{0.25}{1 - 0.25} = 0.333 \text{ kg water/kg d.s.}$$

Step 2
Calculate the time for the constant-rate period:

$$t_c = \frac{m_s}{A}\frac{X_o - X_c}{R_c} = \left(2 \text{ kg d.s./m}^2\right)\frac{(9 - 5)\text{kg water/kg d.s.}}{3 \text{ kg water/m}^2\text{h}} = 2.67 \text{ h}$$

Step 3
Calculate the time for the falling-rate period:

Assumption: Due to a lack of more detailed data, assume that the drying rate in the falling-rate period is directly proportional to the free moisture content.

With this assumption the time for the falling-rate period can be calculated from:

$$t_f = \frac{m_s}{A}\frac{X_1 - X_e}{R_c}\ln\frac{X_c - X_e}{X_f - X_e} =$$
$$= \left(2\text{kg d.s./m}^2\right)\frac{(5 - 0.0333)\text{ kg water/kg d.s.}}{3 \text{ kg water/m}^2\text{h}}\ln\frac{5 - 0.033}{0.333 - 0.033} = 9.29 \text{ h}$$

Step 4
Calculate the total time:

$$t = t_c + t_f = 2.67h + 9.29h = 11.96\,h$$

Example 18.3

Calculate the drying time for a product with $m_s/A = 3\,kgd.s./m^2$ and the following values for the rate of drying at various moisture contents:

X kg water/kg d.s.	R kg water/(kg d.s h m^2)
4.0	20.0
3.8	17.0
3.5	14.0
2.5	8.0
2.0	6.0
1.2	3.0
1.0	1.9
0.9	1.6
0.8	1.3
0.7	1.2
0.6	1.0

Solution

Step 1
Plot the given data:

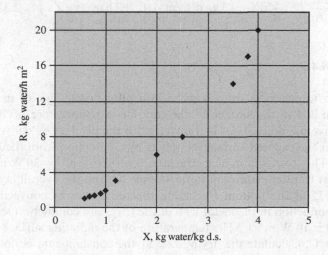

Step 2

As shown in the plot, the relation between R and X is not linear. The drying time can be calculated without the assumption of a linear relation using numerical integration.

By definition R is:

$$R = -\frac{m_s}{A}\frac{dX}{d_t} \qquad (18.1)$$

Integration of eqn (18.1) gives:

$$t = -\frac{m_s}{A}\int_{X_1}^{X_2}\frac{1}{R}dX\frac{m_s}{A}\int_{X_2}^{X_1}\frac{1}{R}dX \qquad (18.2)$$

Since an analytical relationship between 1/R and X is not available, the integral can be approximated using numerical integration. Thus:

$$\int_{X_2}^{X_1}\frac{1}{R}dX \approx \sum\frac{\Delta X}{R_{average}}$$

Numerical integration of $\int_{X_2}^{X_1}\frac{1}{R}dX$ gives (see the spreadsheet program *Drying 2.xls*):

$$\int_{X_2}^{X_1}\frac{1}{R}dX \approx \sum\frac{\Delta X}{R_{average}} = 0.782\,h\,m^2/kg\,d.s.$$

Substitute values in eqn (18.2) and calculate the time:

$$t = \frac{m_s}{A}\int_{X_2}^{X_1}\frac{1}{R}dX = \left(3\,kg\,d.s./m^2\right)\left(0.782\,h\,m^2/kg\,d.s.\right) = 2.35\,h$$

Example 18.4

Tomato paste placed in trays is dried in a pilot dryer. The air in the dryer flowing parallel to the surface of the tray has a temperature of 100 °C and 2.5% relative humidity. Heat is transferred to the product by convection and radiation at the exposed surface as well as by conduction from the bottom of the tray. The convection heat transfer coefficient h_c is 30 W/m² °C, the radiant heat transfer coefficient h_R is 9 W/m² °C, and the overall heat transfer coefficient U at the bottom (including the resistances of convection at the bottom, conduction in the metal wall of the tray, and conduction through the solid food) is 10 W/m² °C. The temperature of the radiating surface above the tray is 110 °C. Calculate the drying rate at the constant-rate period.

Solution

Step 1
Draw the process diagram:

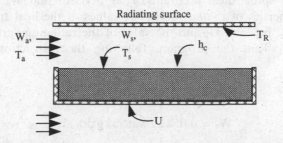

Step 2
Since radiation and conduction are included in this process, the surface temperature of the product T_s will be higher than the wet bulb temperature of the air even in the constant-rate period. The drying rate at the constant-rate period for this case can be calculated from (Ref. 3,7):

$$R_c = \frac{q}{A\lambda_s} = \frac{(h_c + U)(T_a - T_s) + h_R(T_R - T_s)}{\lambda_s} = k_y M_a (W_s - W_a) \quad (18.3)$$

Eqn (18.1) can be rearranged to give:

$$\frac{(W_s - W_a)\lambda_s}{c_s} = \left(1 + \frac{U}{h_c}\right)(T_a - T_s) + \frac{h_R}{h_c}(T_R - T_s) \quad (18.4)$$

Step 3
Find W_s and T_s:

i) Find the humidity and the humid heat of the air from the psychrometric chart or from the spreadsheet program *Moist air properties.xls*:

$$W_a = 0.016 \, \text{kg water/kg dry air}$$
$$c_s = 1.035 \, \text{kJ/kg°C}$$

ii) Substitute values in eqn (18.4):

$$\frac{(W_s - 0.016)\lambda_s}{1.035} = \left(1 + \frac{10}{30}\right)(100 - T_s) + \frac{9}{30}(110 - T_s) \quad (18.5)$$

W_s and T_s represent a point on the saturation humidity curve on the psychrometric chart. Thus, eqn (18.5) has to be solved by trial and error so that W_s and T_s will satisfy eqn (18.5) and the saturation humidity curve on the psychrometric chart. Such a trial and error solution is given in the spreadsheet program *Drying 3.xls*.

iii) Run the spreadsheet program *Drying 3.xls* to find W_s and T_s.Insert the properties of the air, insert the values of the heat transfer coefficients h_c, h_R, and U, and the value of the radiating surface temperature T_R. Run the program, following the instructions. Read the results.

$$\text{Read } T_s = 42.8°\text{C}$$

$$W_s = 0.058 \text{ kg water/kg dry air}$$

Step 4
Substitute values in eqn (18.3) and find R_c:

$$R_c = \frac{(30+10)(100-42.8)+9(110-42.8)}{2400000} = 1.2 \times 10^{-3}\text{kg/m}^2\text{s}$$

$$= 4.34\text{kg/m}^2\text{h}$$

R_c is also given directly in the program *Drying 3.xls*.

Example 18.5

Slices of beets are dried in a dryer as part of a process to produce dehydrated powder for use as a coloring agent in a dressing. The air enters the dryer at a flow rate of 12000 m^3/h, 50 °C temperature, and 10% relative humidity, and exits at 34 °C temperature. If the initial moisture content of the beet is 90% and the beet feed rate is 100 kg/h, calculate the final moisture content of the product.

Solution

Step 1
Draw the process diagram:

Step 2
State your assumptions:
The air in the dryer follows an adiabatic humidification path.

Step 3
Using the psychrometric chart or the spreadsheet program *Moist air properties.xls*, find the air properties at the inlet and the outlet of the dryer.

Inlet: Humidity $W_i = 0.0077$ kg water/kg d.a.
    Humid volume $v_H = 0.925$ m^3/kg d.a.
    Wet bulb temperature $T_w = 23.9°C$

Since the air follows an adiabatic humidification path during its flow through the dryer, the wet bulb temperature at the outlet will be equal to that at the inlet. Thus, using the dry bulb temperature at the exit and the wet bulb temperature as input variables in the psychrometric chart or the spreadsheet program *Moist air properties.xls*, find the humidity at the outlet.

$$\text{Humidity } W_o = 0.0145 \text{ kg water/kg d.a.}$$

Step 4
Calculate the mass flow rate of air:

$$m_a = \frac{Q}{v_H} = \frac{12000 \text{ m}^3/\text{h}}{0.925 \text{ m}^3/\text{kg d.a.}} = 12973 \text{ kg d.a./h}$$

Step 5
Express the product feed rate and the moisture at the inlet on a dry basis:

$$m_a = m_{Fi}(1 - Y_i) = \left(100 \frac{\text{kg product}}{\text{h}}\right)(1 - 0.90) \frac{\text{kg dry solids}}{\text{kg product}} = 10 \frac{\text{kg dry solids}}{\text{h}}$$

$$X_i = \frac{m_{Fi} - m_s}{m_s} = \frac{(100 - 10) \text{kg water/h}}{10 \text{ kg dry solids/h}} = 9 \frac{\text{kg water}}{\text{kg dry solids}}$$

Step 6
Write a moisture balance over the dryer to calculate X_o:

$$m_s X_i + m_a W_i = m_s X_o + m_a W_o$$

Solve for X_o:
$$X_o = \frac{m_s X_i - m_a(W_o - W_i)}{m_s} =$$

$$= \frac{(10 \text{ kg d.s./h})(9 \text{ kg water/kg d.s.}) - (12973 \text{ kg d.a./h})(0.0145 - 0.0077) \text{kg water/kg d.a.}}{10 \text{ kg dry solids/h}} =$$

$$= 0.178 \text{ kg water/kg dry solids}$$

The moisture of the product at the outlet on a wet basis is:

$$Y_o = \frac{X_o}{1 + X_o} = \frac{0.178}{1 + 0.178} \frac{\text{kg water/kg d.s.}}{\text{kg products/kg d.s.}} = 0.151 \text{kg water/kg product}$$

or 15.1 % moisture content

Exercises

Exercise 18.1

A fruit is dried from 85% to 40% moisture (wet basis). Express the moisture on a dry basis and calculate the amount of water that is removed from 1000 kg of product.

Solution

Step 1
Calculate the moisture content on a dry basis:

$$X_{in} = \frac{\text{.........................}}{\text{...........................}} =$$

$$X_{out} = \frac{\text{.........................}}{\text{...........................}} =$$

Step 2
Calculate the amount of water removed:

  i) Calculate the amount of dry solids in 1000 kg of product:

   m_s = ...

  ii) Calculate the amount of water evaporated:

   $m_w = m_s(X_{in} - X_{out}) =$ ....kg water

Exercise 18.2

The following data were obtained in a drying test of a solid food placed in a laboratory tray dryer. Find the critical moisture content, the critical drying rate, the drying time in the constant-rate period, the time necessary for the moisture to fall from the critical moisture content to 70% (wet basis), and the amount of water that was removed from the product in the drying test. It is given that the initial moisture content is 85% wet basis, the bulk density of the solid is 50 kg dry solids /m^3, and the thickness of the solid in the tray is 3 cm.

Time (h)	Weight (kg)
0	3.000
0.5	2.500
1	2.000

(continued)	
Time (h)	Weight (kg)
2	1.600
3	1.500
4	1.435
6	1.420
8	1.410
10	1.405
12	1.400
13	1.400

Solution

Step 1
Express the moisture content of the given data on a dry basis for each given point.

Step 2
Calculate the amount of dry solid per unit surface area:

$$\frac{m_s}{A} = \text{thickness in the tray} \times \ldots\ldots\ldots\ldots = \ldots\ldots\ldots\ldots \times \ldots\ldots\ldots\ldots$$

$$= \ldots\ldots\ldots\ldots \text{kg d.s.}/\text{m}^2$$

Step 3
Calculate the drying rate R for each given point from:

$$R = -\frac{m_s}{A}\frac{dX}{dt}$$

Step 4
Plot R vs X.

Step 5
Read the values of X_c and R_c from the plot.

Step 6
Calculate the drying time in the constant-rate period:

$$t = \frac{}{\ldots\ldots\ldots\ldots\ldots\ldots\ldots\ldots} = \ldots\ldots\ldots\ldots\ldots\ldots\ldots\ldots$$

Step 7
Read the drying rate at a moisture content of 80% and 70% from the plot, and calculate the required drying time using the assumption that the falling rate period is a linear function of X.

$$t = \frac{}{\ldots\ldots\ldots\ldots\ldots\ldots\ldots\ldots} = \ldots\ldots\ldots\ldots\ldots\ldots\ldots\ldots$$

Step 8
Calculate the amount of water removed in the drying test:

$$m_w = \text{.....................................}$$

Exercise 18.3

Using the spreadsheet program *Drying 3.xls*, study the effect of radiation and conduction on the product surface temperature and the drying rate at the constant-rate period for Example 18.4, leaving h_c constant.

Solution

For the following cases, read the drying rate R_c and the surface temperature T_s. Compare T_s to the wet bulb temperature T_w. Comment on the results.

1) $h_r = 0, T_R = 110$ and $U = 0$
2) $h_r = 9, T_R = 110$ and $U = 0$
3) $h_r = 10, T_R = 150$ and $U = 0$
4) $h_r = 0, T_R = 110$ and $U = 10$
5) $h_r = 0, T_R = 110$ and $U = 20$

Exercise 18.4

Diced potatoes initially dried in a belt dryer to 20% moisture will be further dried in a bin dryer to 5% moisture. Calculate the required time to dry a potato cube from 20% to 5% if the effective diffusion coefficient of moisture in potatoes is $2.6 \times 10^{-11} m^2/s$, the equilibrium moisture content at the temperature and humidity of the air used is 0.03 kg water/kg d.s., and the size of the potato cube is 10 mm × 10 mm × 10 mm.

Solution

Step 1
State your assumptions:

- Moisture movement in the potato cube is diffusion controlled.
- Moisture is initially uniform throughout the cube.
- The external resistance to mass transfer is negligible.
- The diffusion coefficient is constant.
- Shrinkage is neglected.

Step 2
Express the moisture content in kg water/kg dry solid:

Initial moisture content $X_o = \text{...}$

Average final moisture content X_m = ...
Equilibrium moisture content X_e = ...

Step 3
Apply the solution of the unsteady state mass transfer equation (Fick's 2nd law) to the average concentration in the x, y, and z directions for negligible surface resistance (see Table 10.1 and Table 13.1):

$$\frac{X_e - X_m}{X_e - X_o} = \frac{8}{\pi^2}\sum_{n=0}^{\infty}\frac{1}{(2n+1)^2}\exp\left(-\frac{(2n+1)^2\pi^2}{4}Fo\right) =$$

$$= \frac{8}{\pi^2}\left(\exp(-2.47Fo) + \frac{1}{9}\exp(-22.2Fo) + \frac{1}{25}\exp(-61.7Fo) + ...\right)$$

Use the spreadsheet program *Mass transfer-Negligible Surface Resistance.xls*. Insert the parameter values L_x, L_y, L_z, C_o, C_e and D_{ij}.
Run the program until the $C_{mean} = X_m$. Read the time.

Exercise 18.5

In a vegetable dryer, room air at 25 °C and 60% relative humidity is mixed with a portion of the exit air from the dryer. The mixture is heated to 90 °C and attains 10% relative humidity before entering the dryer. The air leaves the dryer at 60 °C. Find the relative humidity of the air at the exit of the dryer, the required fresh air flow rate, the recirculation rate, and the steam consumption in the air heater, if the initial moisture content of the vegetable is 80%, the final moisture content is 20% (wet basis), the feed rate to the dryer is 500 kg/h, and the air is indirectly heated by 120 °C saturated steam. If, for the same product conditions, air recirculation was not used, but instead the same total mass air flow rate was heated to 90 °C before entering the dryer, calculate the steam consumption and the relative humidity of the air at the exit.

Solution

Step 1
Draw the process diagram:

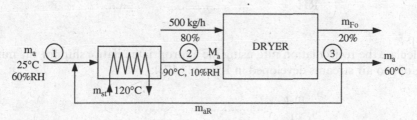

Step 2
State your assumptions: The operation of the dryer is adiabatic.

Step 3
Calculate the amount of water evaporated in the dryer:

i) Calculate the dry solids content in the feed:

$$m_s = \ldots\ldots\ldots\ldots\ldots\ldots\ldots\ldots\ldots\ldots\ldots$$

ii) Calculate the moisture content of the product at the inlet and the outlet on a dry basis:

$$X_i = \frac{\rule{3cm}{0.4pt}}{\ldots\ldots\ldots\ldots\ldots} = \ldots\ldots\ldots\ldots\text{kg water/kg d.s.}$$

$$X_o = \frac{\rule{3cm}{0.4pt}}{\ldots\ldots\ldots\ldots\ldots} = \ldots\ldots\ldots\ldots\text{kg water/kg d.s.}$$

iii) Calculate the amount of water evaporated:

$$m_w = \ldots\ldots\ldots \frac{\text{kg d.s.}}{h} (\ldots\ldots - \ldots\ldots) \frac{\text{kg water}}{\text{kg d.s.}} = \ldots\ldots \frac{\text{kg water}}{h}$$

Step 4
Find the humidity and the enthalpy at points 1, 2, and 3 from the psychrometric chart or from the spreadsheet program *Moist air properties.xls*, as well as the relative humidity at point 3:

$$W_1 = \ldots\ldots\ldots\ldots\ldots\ldots\ldots\ldots$$
$$W_2 = \ldots\ldots\ldots\ldots\ldots\ldots\ldots\ldots$$
$$W_3 = \ldots\ldots\ldots\ldots\ldots\ldots\ldots\ldots$$
$$H_1 = \ldots\ldots\ldots\ldots\ldots\ldots\ldots\ldots$$
$$H_2 = \ldots\ldots\ldots\ldots\ldots\ldots\ldots\ldots$$
$$H_3 = \ldots\ldots\ldots\ldots\ldots\ldots\ldots\ldots$$
$$RH_3 = \ldots\ldots\ldots\ldots\ldots\ldots\ldots\ldots$$

Step 5
Calculate the recirculation rate using the appropriate relationship for the mixing of two air streams developed in Example 17.4:

$$\frac{m_{aR}}{m_2} = \ldots\ldots\ldots\ldots\ldots\ldots\ldots\ldots$$

Step 6
Calculate the mass flow rate of air in the dryer by dividing the total amount of water evaporated, by the moisture picked up by the air:

$$m_2 = \text{...kg dry air/h}$$

Step 7
Calculate the mass flow rate of fresh air to the heater from an air mass balance on the heater and the known ratio of $\frac{m_{aR}}{m_2}$:

$$m_a = \text{...}$$

Step 8
Find the humid volume of fresh air and calculate its volumetric flow rate:

$$Q_a = \text{...m}^3/\text{h}$$

Step 9
Write an enthalpy balance in the air heater to calculate the steam consumption (assume that the condensate exits at the saturation temperature):

$$m_{st} = \text{...kg/h}$$

Step 10
Draw the process diagram for the case in which recirculation is not used:

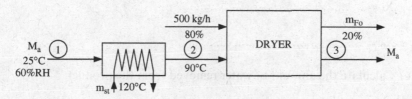

Step 11
Find the enthalpy of the air at point 2 from the psychrometric chart or from the spreadsheet program *Moist air properties.xls*, and write an enthalpy balance in the air heater to calculate the steam consumption.

Step 12
Write a moisture balance on the dryer and calculate the moisture content of the air at point 3.

Step 13
Find the relative humidity at point 3 from the psychrometric chart or from the spreadsheet program *Moist air properties.xls*.

Step 14
Compare the results.

Exercise 18.6

A countercurrent tunnel dryer is used to dry 500 kg/h of a vegetable from 85% to 15% moisture. Room air at 20 °C and 50% relative humidity is heated to 90 °C in an air heater before entering the dryer, flows through the dryer, and exits at 90% relative humidity. Calculate the required air flow rate and the required energy in the air heater.

Solution

Step 1
Draw the process diagram.

Step 2
State your assumptions: ....

Step 3
Calculate the amount of water removed from the product.

   a) Calculate the solids content:

   $X_s = $ ....

   b) Calculate moisture content of the product on a dry basis at the inlet and the outlet of the dryer:

   $X_i = $ ....

   $X_0 = $ ....

   c) Calculate the amount of water removed from the product:

   $m_w = $ ....kg/h

Step 4
Find the air properties at the inlet:

$$W_1 = \text{................................}$$

$$v_{H1} = \text{................................}$$

$$H_1 = \text{................................}$$

Step 5
Find the air properties after the heater:

$$T_{w2} = \text{................................}$$

$$H_2 = \text{................................}$$

Step 6
Find the air properties at the exit of the dryer:

$$W_3 = \text{................................}$$

$$T_{w3} = \text{................................}$$

Step 7
Use m_w, W_1, and W_3 to calculate the required air flow rate:

$$m_a = \text{................................}$$

Step 8
Use m_a and $v_{H\,1}$ to calculate the volumetric air flow rate at the inlet:

$$Q_a = \text{................................}$$

Step 9
Use m_a, H_1, and H_2 to calculate the required energy input to heat the air:

$$q = \text{................................}$$

Exercise 18.7

If the inlet air conditions in the dryer of the previous problem change to 20 °C and 90% relative humidity due to a sudden weather change, calculate the air flow rate if the moisture of the product at the exit of the dryer, the product flow rate, the temperature of the air after the heater, and the relative humidity of the air at the exit remain the same as in the previous problem. Calculate the ratio of the product flow rate to the air flow rate and compare it with that of the previous problem.

Solution

Hint: Follow similar reasoning to that of the previous problem.

Exercise 18.8

For the production of skim milk powder, the milk is preheated, centrifuged to separate the cream, pasteurized, and concentrated in a multiple effect evaporator. Following these processing steps, concentrated skim milk at a rate of 5000 kg/h with 50% moisture will be spray dried to 4% moisture in a spray dryer where room air at 20 °C and 60% relative humidity is heated to 200 °C before entering the dryer. To provide an adequate driving force for mass transfer and reach the required final moisture content of the product, it is necessary to have the relative humidity of the air at the exit equal to 10%. Calculate the required air flow rate, the temperature of the air at the exit, the heat required, and the specific heat consumption (heat required per kg water evaporated).

Solution

Hint: Follow similar reasoning to that of the previous problem.

Exercise 18.9

Use the spreadsheet program *Continuous dryer.xls* to solve Exercises 18.6, 18.7, and 18.8.

References

1 Dodge D.W. and Metzner A.B. (1959). Turbulent flow of non-Newtonian systems. *A.I. Ch.E. J.* 5, 189-204.
2 Singh P.R. and Heldman D.R. (1993). *Introduction to Food Engineering*, 2nd Edition, Academic Press, London.
3 Geankoplis C. J. (1978). *Transport Processes and Unit Operations*. Allyn & Bacon, Boston, Massachusetts.
4 Inropera F.P. and De Witt D.P. (1990). *Fundamentals of Heat and Mass Transfer*, 3rd Edition. Jon Wiley & Sons, New York.
5 Crank J. (1975). *The Mathematics of Diffusion*, 2nd Edition, Oxford Science Publications, Clarendon Press, Oxford.
6 Carlsaw H.S. and Jaeger J.C. (1959). *Conduction of Heat in Solids*. Oxford, Clarendon Press
7 Treybal R.E. (1968). *Mass Transfer Operations*. McGraw-Hill Book Company, New York.
8 American Society of Heating, Refrigeration and Air Conditioning Engineers, Inc. (1998). "1998 ASHRAE Handbook, Refrigeration", p. 8.2.
9 Heldman D.R. (1992). "Food Freezing" in *Handbook of Food Engineering*. Edited by D. Heldman and D. Lund. Marcel Dekker, Inc., New York.
10 Cleland D.J. and Valentas K.J. (1997). "Prediction of Freezing Time and Design of Food Freezers" in *Handbook of Food Engineering Practice*, Edited by K. Valentas, E. Rotsrein, and R.P. Singh. CRC Press, New York.
11 Schwartzberg H., Singh P.R., and Sarkar A. (2007). "Freezing and Thawing of Foods" in *Heat Transfer in Food Processing*. Edited by S. Yanniotis & B. Sunden. WIT Press, UK.
12 Villota R. and Hawkes J.G. (1992). "Reaction Kinetics in Food Systems" in *Handbook of Food Engineering*. Edited by D. Heldman and D. Lund. Marcel Dekker, Inc., New York.
13 Ramaswamy H.S. and Singh P.R. (1997). "Sterilization Process Engineering" in *Handbook of Food Engineering Practice*. Edited by K. Valentas, E. Rotsrein, and R.P. Singh. CRC Press, New York.
14 Stadelman, W.J., V.M.Olson, G.A. Shemwell, and S. Pasch (1988). *Egg and Meat Processing*. Ellis Harwood, Ltd., Chickester, England.
15 American Society of Heating, Refrigeration and Air Conditioning Engineers, Inc. (1977). "ASHRAE Handbook, 1977 Fundamentals", p. 5.2.

Appendix

Appendix: Answers to Review Questions

Chapter 2		Chapter 3		Chapter 4		Chapter 5	
1	T	1	T	1	T	1	T
2	T	2	T	2	T	2	T
3	T	3	T	3	T	3	T
4	F	4	T	4	T	4	T
5	T	5	F	5	T	5	T
6	F	6	F	6	F	6	F
7	T	7	T	7	T	7	T
8	T	8	T	8	F	8	T
9	T	9	T	9	T	9	F
10	F	10	F	10	T	10	F
11	T					11	T
12	F					12	T
13	F					13	T
14	T					14	F
15	T					15	T
16	T					16	T
17	T					17	F
18	T					18	T
19	F					19	F
20	T					20	T
21	T						
22	T						
23	T						
24	F						
25	T						

T = True, F = False

Chapter 6		Chapter 7		Chapter 8		Chapter 9	
1	T	1	T	1	T	1	T
2	F	2	T	2	T	2	F
3	F	3	F	3	F	3	T
4	T	4	T	4	T	4	T
5	F	5	T	5	F	5	F
6	T	6	T	6	F	6	T
7	F	7	F	7	T	7	F
8	T	8	T	8	T	8	F
9	T	9	F	9	T	9	T
10	T	10	F	10	T	10	T
11	F	11	T	11	T	11	T
12	T	12	T	12	F	12	F
13	T	13	F	13	T	13	T
14	T	14	F	14	T	14	F
15	T	15	T	15	T	15	F
16	T	16	T	16	F		
17	T	17	T	17	T		
18	T	18	F	18	T		
19	F	19	F	19	T		
20	T	20	T	20	T		
21	T			21	F		
22	F			22	T		
23	T			23	F		
24	T			24	T		
25	F			25	T		
26	F			26	F		
27	T			27	T		
28	T			28	F		
29	F			29	T		
30	T			30	F		

Chapter 10		Chapter 11		Chapter 12		Chapter 13	
1	T	1	T	1	F	1	T
2	T	2	T	2	T	2	T
3	F	3	F	3	T	3	T
4	T	4	T	4	F	4	F
5	F	5	T	5	T	5	F
6	T	6	F	6	T	6	T
7	F	7	T	7	F	7	T
8	T	8	T	8	T	8	F
9	T	9	F	9	T	9	F
10	F	10	T	10	T	10	T
11	T	11	F				
12	T	12	T				
13	F	13	T				
14	T	14	T				
15	T	15	T				
		16	T				
		17	F				
		18	T				
		19	T				
		20	T				

Chapter 14		Chapter 15		Chapter 16			
1	T	1	T	1	T	31	T
2	T	2	F	2	T	32	T
3	T	3	F	3	F	33	T
4	F	4	T	4	T	34	T
5	T	5	T	5	T	35	T
6	F	6	T	6	T	36	T
7	T	7	T	7	F	37	T
8	T	8	T	8	T	38	F
9	F	9	T	9	F	39	T
10	F	10	T	10	F	40	T
11	F	11	F	11	T	41	F
12	T	12	T	12	T	42	F
13	T	13	F	13	F	43	T
14	T	14	T	14	F	44	T
15	T	15	T	15	T	45	F
16	T	16	T	16	T	46	T
17	F	17	T	17	T	47	T
18	T	18	F	18	T	48	T
19	F	19	F	19	T	49	F
20	T	20	T	20	F	50	T
21	T	21	T	21	T	51	T
22	F	22	T	22	T	52	T
23	F	23	F	23	F	53	F
24	F	24	T	24	T	54	T
25	T	25	T	25	T	55	T
		26	T	26	T		
		27	F	27	T		
		28	T	28	T		
		29	T	29	T		
		30	F	30	T		

Chapter 17		Chapter 18			
1	T	1	T	26	T
2	F	2	T	27	F
3	F	3	F	28	T
4	F	4	F	29	T
5	F	5	T	30	F
6	F	6	T	31	F
7	T	7	F	32	T
8	T	8	F	33	T
9	T	9	F	34	T
10	T	10	F	35	F
11	T	11	T	36	F
12	T	12	F	37	F
13	F	13	T	38	T
14	T	14	T	39	T
15	T	15	F	40	T
16	T	16	F	41	T
17	T	17	T	42	T
18	T	18	T	43	T
19	T	19	T	44	T
20	T	20	F	45	T
		21	T	46	F
		22	F	47	T
		23	F	48	T
		24	F	49	T
		25	F	50	T

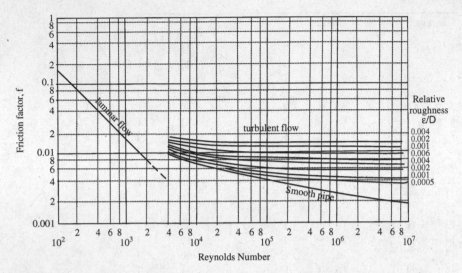

Fig. A.1 Moody diagram of the friction factor for fluid flow inside pipes (equivalent roughness ε, in m, for drawn tubing 1.5×10^{-6}, commercial steel 46×10^{-6}, galvanized iron 150×10^{-6}, cast iron 260×10^{-6}) Based on Moody LF (1944) T ASME 66. Used with permission

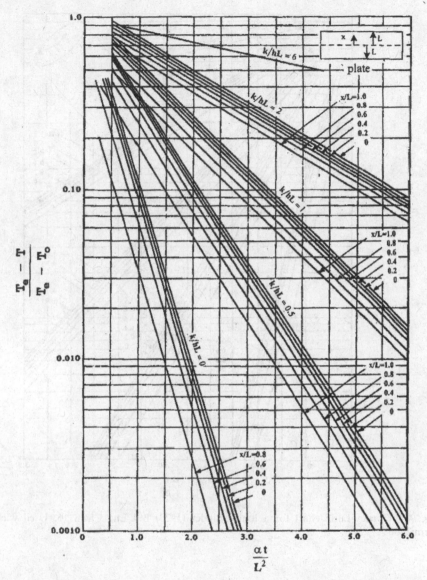

Fig. A.2 Gurney-Lurie chart for a flat plate (1923) Ind Eng Chem 15. Used with permission

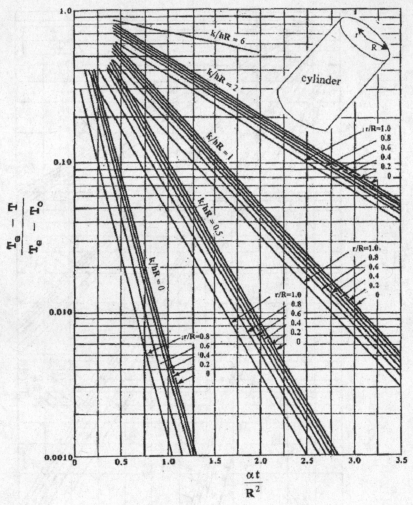

Fig. A.3 Gurney-Lurie chart for a long cylinder (1923) Ind Eng Chem 15. Used with permission

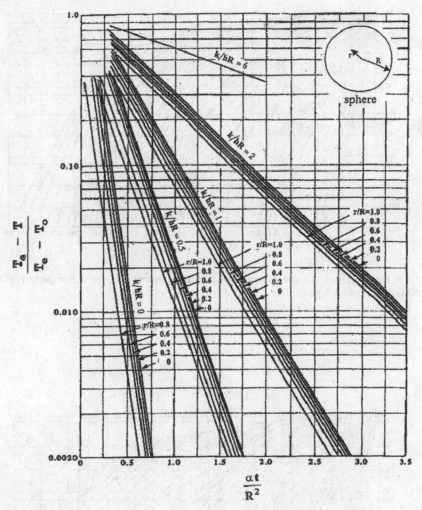

Fig. A.4 Gurney-Lurie chart for a sphere (1923) Ind Eng Chem 15. Used with permission

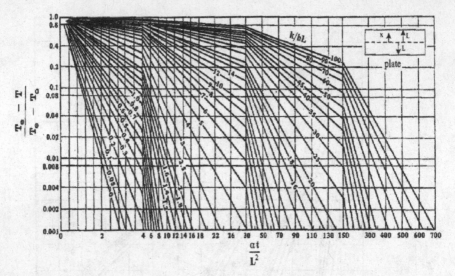

Fig. A.5 Heisler chart for determining the midplane temperature of a flat plate (1947) T ASME 69. Used with permission

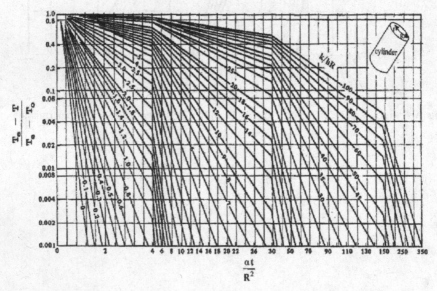

Fig. A.6 Heisler chart for determining the centerline temperature of a long cylinder (1947) T ASME 69. Used with permission

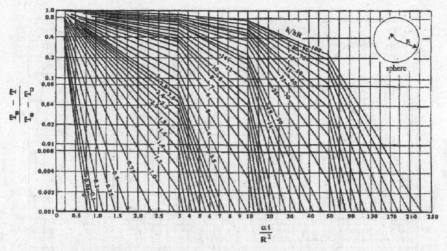

Fig. A.7 Heisler chart for determining the center temperature of a sphere (1947) T ASME 69. Used with permission

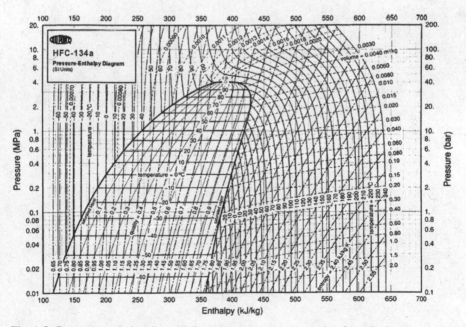

Fig. A.8 Pressure- enthalpy diagram for HFC 134a with permission from DuPont

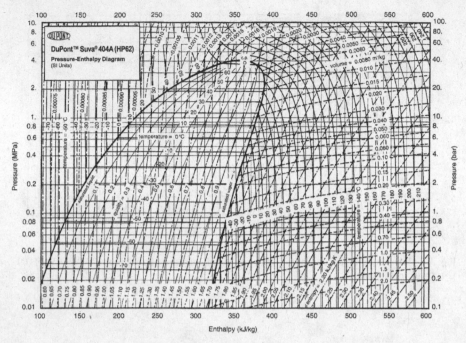

Fig. A.9 Pressure- enthalpy diagram for HFC 404a with permission from DuPont

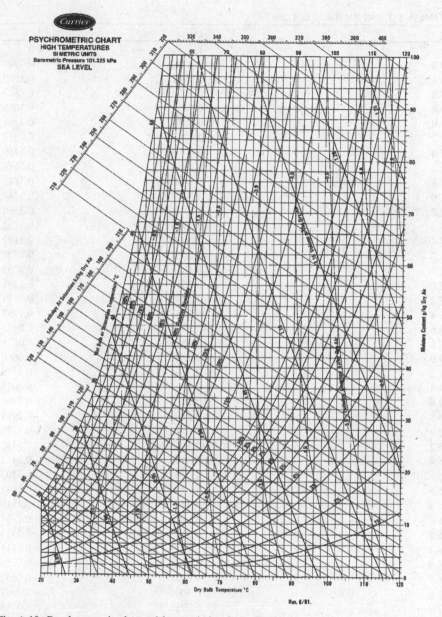

Fig. A.10 Psychrometric chart with permission from Carrier

Table A.1 Bessel functions $J_0(x)$ and $J_1(x)$

x	$J_0(x)$	$J_1(x)$
0	1	0
0.2	0.9900	0.0995
0.4	0.9604	0.1960
0.6	0.9120	0.2867
0.8	0.8463	0.3688
1.0	0.7652	0.4401
1.2	0.6711	0.4983
1.4	0.5669	0.5419
1.6	0.4554	0.5699
1.8	0.3400	0.5815
2.0	0.2239	0.5767
2.2	0.1104	0.5560
2.4	0.0025	0.5202
2.4048	0.0000	0.5192
2.6	−0.0968	0.4708
2.8	−0.1850	0.4097
3.0	−0.2601	0.3391
3.2	−0.3202	0.2613
3.4	−0.3643	0.1792
3.6	−0.3918	0.0955
3.8	−0.4026	0.0128
4.0	−0.3971	−0.0660
4.2	−0.3766	−0.1386
4.4	−0.3423	−0.2028
4.6	−0.2961	−0.2566
4.8	−0.2404	−0.2985
5.0	−0.1776	−0.3276
5.2	−0.1103	−0.3432
5.4	−0.0412	−0.3453
5.5201	0.0000	−0.3403
5.6	0.0270	−0.3343
5.8	0.0917	−0.3110
6.0	0.1506	−0.2767
6.2	0.2017	−0.2329
6.4	0.2433	−0.1816
6.6	0.2740	−0.1250
6.8	0.2931	−0.0652
7.0	0.3001	−0.0047
7.2	0.2951	0.0543
7.4	0.2786	0.1096
7.6	0.2516	0.1592
7.8	0.2154	0.2014
8.0	0.1717	0.2346
8.2	0.1222	0.2580
8.4	0.0692	0.2708
8.6	0.0146	0.2728

Table A.1 (continued)

x	$J_0(x)$	$J_1(x)$
8.6537	0.0000	0.2715
8.8	−0.0392	0.2641
9.0	−0.0903	0.2453
9.2	−0.1367	0.2174
9.4	−0.1768	0.1816
9.6	−0.2090	0.1395
9.8	−0.2323	0.0928
10.0	−0.2459	0.0435
10.2	−0.2496	−0.0066
10.4	−0.2434	−0.0555
10.6	−0.2276	−0.1012
10.8	−0.2032	−0.1422
11.0	−0.1712	−0.1768
11.2	−0.1330	−0.2039
11.4	−0.0902	−0.2225
11.6	−0.0446	−0.2320
11.7915	0.0000	−0.2325
11.8	0.0020	−0.2323
12.0	0.0477	−0.2234
12.2	0.0908	−0.2060
12.4	0.1296	−0.1807
12.6	0.1626	−0.1487
12.8	0.1887	−0.1114
13.0	0.2069	−0.0703
13.2	0.2167	−0.0271
13.4	0.2177	0.0166
13.6	0.2101	0.0590
13.8	0.1943	0.0984
14.0	0.1711	0.1334
14.2	0.1414	0.1626
14.4	0.1065	0.1850
14.6	0.0679	0.1999
14.8	0.0271	0.2066
14.9309	0.0000	0.2065
15.0	−0.0142	0.2051
16.0	−0.1749	0.0904
17.0	−0.1699	−0.0977
18.0	−0.0134	−0.1880
18.0711	0.0000	−0.1877
20.0	0.1670	0.0668

Table A.2 First six roots of the equation: $\delta \tan \delta = Bi$

Bi	δ_1	δ_2	δ_3	δ_4	δ_5	δ_6
0	0	3.1416	6.2832	9.4248	12.5664	15.7080
0.10	0.3111	3.1731	6.2991	9.4354	12.5743	15.7143
0.15	0.3779	3.1886	6.3070	9.4407	12.5783	15.7175
0.20	0.4328	3.2039	6.3148	9.4459	12.5823	15.7207
0.25	0.4792	3.2191	6.3226	9.4512	12.5862	15.7239
0.30	0.5218	3.2341	6.3305	9.4565	12.5902	15.7270
0.35	0.5591	3.2489	6.3383	9.4618	12.5942	15.7302
0.40	0.5932	3.2636	6.3461	9.4670	12.5981	15.7334
0.45	0.6248	3.2780	6.3539	9.4722	12.6021	15.7365
0.50	0.6533	3.2923	6.3616	9.4775	12.6060	15.7397
0.55	0.6798	3.3064	6.3693	9.4827	12.6100	15.7429
0.60	0.7051	3.3204	6.3770	9.4879	12.6139	15.7460
0.65	0.7283	3.3341	6.3846	9.4931	12.6178	15.7492
0.70	0.7506	3.3477	6.3923	9.4983	12.6218	15.7524
0.75	0.7713	3.3611	6.3998	9.5035	12.6257	15.7555
0.80	0.7910	3.3744	6.4074	9.5087	12.6296	15.7587
0.85	0.8096	3.3873	6.4149	9.5139	12.6335	15.7618
0.90	0.8274	3.4003	6.4224	9.5190	12.6375	15.7650
0.95	0.8442	3.4130	6.4299	9.5242	12.6414	15.7681
1.0	0.8603	3.4256	6.4373	9.5293	12.6453	15.7713
1.2	0.9181	3.4742	6.4667	9.5498	12.6609	15.7839
1.4	0.9663	3.5201	6.4955	9.5700	12.6764	15.7964
1.6	1.0083	3.5636	6.5237	9.5901	12.6918	15.8088
1.8	1.0448	3.6049	6.5513	9.6099	12.7071	15.8212
2.0	1.0769	3.6436	6.5783	9.6296	12.7223	15.8336
2.2	1.1052	3.6803	6.6047	9.6489	12.7374	15.8459
2.4	1.1305	3.7151	6.6305	9.6681	12.7524	15.8581
2.6	1.1533	3.7480	6.6556	9.6870	12.7673	15.8703
2.8	1.1738	3.7792	6.6801	9.7056	12.7820	15.8824
3.0	1.1925	3.8088	6.7040	9.7240	12.7967	15.8945
3.5	1.2323	3.8761	6.7609	9.7688	12.8326	15.9243
4.0	1.2646	3.9352	6.8140	9.8119	12.8678	15.9536
4.5	1.2913	3.9873	6.8635	9.8532	12.9020	15.9824
5.0	1.3138	4.0336	6.9096	9.8928	12.9352	16.0107
6.0	1.3496	4.1116	6.9924	9.9667	12.9988	16.0654
7.0	1.3766	4.1746	7.0640	10.0339	13.0584	16.1177
8.0	1.3978	4.2264	7.1263	10.0949	13.1141	16.1675
9.0	1.4149	4.2694	7.1806	10.1502	13.1660	16.2147
10	1.4289	4.3058	7.2281	10.2003	13.2142	16.2594
12	1.4505	4.3636	7.3070	10.2869	13.3004	16.3414
14	1.4664	4.4074	7.3694	10.3586	13.3746	16.4142
16	1.4786	4.4416	7.4198	10.4184	13.4386	16.4786
18	1.4883	4.4690	7.4610	10.4688	13.4939	16.5357
20	1.4961	4.4915	7.4954	10.5117	13.5420	16.5864
25	1.5105	4.5330	7.5603	10.5947	13.6378	16.6901
30	1.5202	4.5615	7.6057	10.6543	13.7085	16.7691
35	1.5272	4.5822	7.6391	10.6989	13.7625	16.8305
40	1.5325	4.5979	7.6647	10.7334	13.8048	16.8794

Table A.3 First six roots of the equation: $\delta J_1(\delta) - Bi\, J_0(\delta) = 0$

Bi	δ_1	δ_2	δ_3	δ_4	δ_5	δ_6
0	0.0348	3.8317	7.0156	10.1735	13.3237	16.4706
0.10	0.4421	3.8577	7.0298	10.1833	13.3312	16.4767
0.15	0.5375	3.8706	7.0369	10.1882	13.3349	16.4797
0.20	0.6181	3.8835	7.0440	10.1931	13.3387	16.4828
0.25	0.6861	3.8963	7.0511	10.1980	13.3424	16.4858
0.30	0.7464	3.9090	7.0582	10.2029	13.3462	16.4888
0.35	0.8026	3.9217	7.0652	10.2078	13.3499	16.4919
0.40	0.8524	3.9343	7.0723	10.2127	13.3537	16.4949
0.45	0.8984	3.9469	7.0793	10.2176	13.3574	16.4979
0.50	0.9412	3.9593	7.0864	10.2224	13.3611	16.5009
0.55	0.9812	3.9716	7.0934	10.2272	13.3649	16.5039
0.60	1.0184	3.9840	7.1003	10.2322	13.3686	16.5070
0.65	1.0538	3.9963	7.1073	10.2370	13.3723	16.5100
0.70	1.0872	4.0084	7.1143	10.2419	13.3761	16.5130
0.75	1.1196	4.0205	7.1213	10.2468	13.3798	16.5161
0.80	1.1495	4.0324	7.1282	10.2516	13.3835	16.5191
0.85	1.1780	4.0443	7.1351	10.2565	13.3872	16.5221
0.90	1.2049	4.0561	7.1420	10.2613	13.3910	16.5251
0.95	1.2309	4.0678	7.1489	10.2661	13.3947	16.5281
1.0	1.2558	4.0795	7.1557	10.2709	13.3984	16.5311
1.2	1.3457	4.1249	7.1830	10.2902	13.4132	16.5432
1.4	1.4226	4.1689	7.2099	10.3093	13.4279	16.5552
1.6	1.4892	4.2112	7.2364	10.3283	13.4427	16.5672
1.8	1.5476	4.2519	7.2626	10.3471	13.4573	16.5791
2.0	1.5991	4.2910	7.2884	10.3657	13.4718	16.5910
2.2	1.6451	4.3287	7.3136	10.3843	13.4863	16.6029
2.4	1.6868	4.3644	7.3385	10.4026	13.5007	16.6147
2.6	1.7241	4.3988	7.3629	10.4208	13.5151	16.6265
2.8	1.7579	4.4318	7.3868	10.4388	13.5293	16.6382
3.0	1.7884	4.4634	7.4103	10.4565	13.5433	16.6497
3.5	1.8545	4.5364	7.4671	10.5001	13.5782	16.6786
4.0	1.9078	4.6019	7.5200	10.5422	13.6124	16.7072
4.5	1.9523	4.6604	7.5704	10.5829	13.6459	16.7354
5.0	1.9898	4.7133	7.6178	10.6222	13.6784	16.7629
6.0	2.0490	4.8033	7.7039	10.6963	13.7413	16.8167
7.0	2.0935	4.8771	7.7797	10.7645	13.8007	16.8684
8.0	2.1285	4.9384	7.8464	10.8270	13.8566	16.9178
9.0	2.1566	4.9897	7.9051	10.8842	13.9090	16.9650
10	2.1795	5.0332	7.9569	10.9363	13.9580	17.0099
12	2.2147	5.1027	8.0437	11.0274	14.0464	17.0927
14	2.2405	5.1555	8.1128	11.1035	14.1232	17.1669
16	2.2601	5.1967	8.1689	11.1674	14.1898	17.2329
18	2.2756	5.2298	8.2150	11.2215	14.2478	17.2918
20	2.2880	5.2568	8.2534	11.2677	14.2983	17.3441
25	2.3108	5.3068	8.3262	11.3575	14.3996	17.4522
30	2.3262	5.3410	8.3771	11.4222	14.4749	17.5349
35	2.3372	5.3659	8.4146	11.4706	14.5323	17.5994
40	2.3455	5.3847	8.4432	11.5081	14.5774	17.6508

Table A.4 First six roots of the equation: $\delta \cot \delta = 1\text{-}Bi$

Bi	δ_1	δ_2	δ_3	δ_4	δ_5	δ_6
0	0	4.4934	7.7252	10.9041	14.0662	17.2207
0.10	0.5423	4.5157	7.7382	10.9133	14.0733	17.2266
0.15	0.6609	4.5268	7.7447	10.9179	14.0769	17.2295
0.20	0.7593	4.5379	7.7511	10.9225	14.0804	17.2324
0.25	0.8453	4.5490	7.7577	10.9270	14.0839	17.2353
0.30	0.9208	4.5601	7.7641	10.9316	14.0875	17.2382
0.35	0.9895	4.5711	7.7705	10.9362	14.0910	17.2411
0.40	1.0528	4.5822	7.7770	10.9408	14.0946	17.2440
0.45	1.1121	4.5932	7.7834	10.9453	14.0981	17.2469
0.50	1.1656	4.6042	7.7899	10.9499	14.1017	17.2498
0.55	1.2161	4.6151	7.7963	10.9544	14.1053	17.2527
0.60	1.2644	4.6261	7.8028	10.9591	14.1088	17.2556
0.65	1.3094	4.6370	7.8091	10.9636	14.1124	17.2585
0.70	1.3525	4.6479	7.8156	10.9682	14.1159	17.2614
0.75	1.3931	4.6587	7.8219	10.9727	14.1195	17.2643
0.80	1.4320	4.6696	7.8284	10.9773	14.1230	17.2672
0.85	1.3931	4.6587	7.8219	10.9727	14.1195	17.2643
0.90	1.5044	4.6911	7.8412	10.9865	14.1301	17.2730
0.95	1.5383	4.7017	7.8476	10.9910	14.1336	17.2759
1.0	1.5708	4.7124	7.8540	10.9956	14.1372	17.2788
1.2	1.6887	4.7544	7.8794	11.0137	14.1513	17.2903
1.4	1.7906	4.7954	7.9045	11.0318	14.1654	17.3019
1.6	1.8797	4.8356	7.9295	11.0498	14.1795	17.3134
1.8	1.9586	4.8750	7.9542	11.0677	14.1935	17.3249
2.0	2.0288	4.9132	7.9787	11.0855	14.2074	17.3364
2.2	2.0916	4.9502	8.0028	11.1033	14.2213	17.3478
2.4	2.1483	4.9861	8.0267	11.1208	14.2352	17.3592
2.6	2.1996	5.0209	8.0502	11.1382	14.2490	17.3706
2.8	2.2463	5.0545	8.0733	11.1555	14.2627	17.3819
3.0	2.2889	5.0870	8.0962	11.1727	14.2763	17.3932
3.5	2.3806	5.1633	8.1516	11.2149	14.3101	17.4213
4.0	2.4556	5.2329	8.2045	11.2560	14.3433	17.4490
4.5	2.5180	5.2963	8.2550	11.2960	14.3760	17.4764
5.0	2.5704	5.3540	8.3029	11.3348	14.4079	17.5034
6.0	2.6536	5.4544	8.3913	11.4086	14.4699	17.5562
7.0	2.7165	5.5378	8.4703	11.4772	14.5288	17.6072
8.0	2.7654	5.6077	8.5406	11.5408	14.5847	17.6562
9.0	2.8044	5.6669	8.6030	11.5994	14.6374	17.7032
10	2.8363	5.7172	8.6587	11.6532	14.6869	17.7481
12	2.8851	5.7981	8.7527	11.7481	14.7771	17.8315
14	2.9206	5.8597	8.8282	11.8281	14.8560	17.9067
16	2.9476	5.9080	8.8898	11.8959	14.9251	17.9742
18	2.9687	5.9467	8.9406	11.9535	14.9855	18.0346
20	2.9857	5.9783	8.9831	12.0029	15.0384	18.0887
25	3.0166	6.0368	9.0637	12.0994	15.1450	18.2007
30	3.0372	6.0766	9.1201	12.1691	15.2245	18.2870
35	3.0521	6.1055	9.1616	12.2213	15.2855	18.3545
40	3.0632	6.1273	9.1933	12.2618	15.3334	18.4085

Table A.5 Error function

x	erf (x)	erfc (x)
0	0	1
0.05	0.056372	0.943628
0.10	0.112463	0.887537
0.15	0.167996	0.832004
0.20	0.222703	0.777297
0.25	0.276326	0.723674
0.30	0.328627	0.671373
0.35	0.379382	0.620618
0.40	0.428392	0.571608
0.45	0.475482	0.524518
0.50	0.520500	0.479500
0.55	0.563323	0.436677
0.60	0.603856	0.396144
0.65	0.642029	0.357971
0.70	0.677801	0.322199
0.75	0.711155	0.288845
0.80	0.742101	0.257899
0.85	0.770668	0.229332
0.90	0.796908	0.203092
0.95	0.820891	0.179109
1.0	0.842701	0.157299
1.1	0.880205	0.119795
1.2	0.910314	0.089686
1.3	0.934008	0.065992
1.4	0.952285	0.047715
1.5	0.966105	0.033895
1.6	0.976348	0.023652
1.7	0.983790	0.016210
1.8	0.989091	0.010909
1.9	0.992790	0.007210
2.0	0.995322	0.004678
2.1	0.997021	0.002979
2.2	0.998137	0.001863
2.3	0.998857	0.001143
2.4	0.999311	0.000689
2.5	0.999593	0.000407
2.6	0.999764	0.000236
2.7	0.999866	0.000134
2.8	0.999925	0.000075
2.9	0.999959	0.000041
3.0	0.999978	0.000022

Index